# Artificial intelligence and the future of warfare

Manchester University Press

# Artificial intelligence and the future of warfare

## The USA, China, and strategic stability

James Johnson

MANCHESTER UNIVERSITY PRESS

Copyright © James Johnson 2021

The right of James Johnson to be identified as the author of this work has been asserted by them in accordance with the Copyright, Designs and Patents Act 1988.

Published by Manchester University Press
Oxford Road, Manchester M13 9PL

www.manchesteruniversitypress.co.uk

British Library Cataloguing-in-Publication Data
A catalogue record for this book is available from the British Library

ISBN 978 1 5261 4505 5 hardback
ISBN 978 1 5261 7908 1 paperback

First published 2021

The publisher has no responsibility for the persistence or accuracy of URLs for any external or third-party internet websites referred to in this book, and does not guarantee that any content on such websites is, or will remain, accurate or appropriate.

Typeset by
Servis Filmsetting Ltd, Stockport, Cheshire

# Contents

| | |
|---|---|
| List of figures | *page* vi |
| Acknowledgments | vii |
| List of abbreviations | ix |
| Introduction: opening the AI Pandora's box | 1 |

### Part I: Destabilizing the AI renaissance

| | |
|---|---|
| 1  Military AI primer | 17 |
| 2  AI in the second nuclear age | 41 |

### Part II: Military AI superpowers

| | |
|---|---|
| 3  New challenges to military-techno *Pax Americana* | 59 |
| 4  US–China crisis stability under the AI nuclear shadow | 85 |

### Part III: Nuclear instability redux?

| | |
|---|---|
| 5  Hunting for nuclear weapons in the digital age | 111 |
| 6  The fast and the furious: drone swarming and hypersonic weapons | 128 |
| 7  The AI-cyber security nexus | 150 |
| 8  Delegating strategic decisions to intelligent machines | 168 |
| Conclusion: managing an AI future | 198 |
| Index | 221 |

# List of figures

1.1 Major research fields and disciplines associated with AI  *page* 19
1.2 The linkages between AI & autonomy  20
1.3 Hierarchical relationship: ML is a subset of AI, and DL is a subset of ML  22
1.4 The emerging 'AI ecosystem'  23

# Acknowledgments

So, you're writing a book on artificial intelligence (AI); I have to ask you, do you agree with Steven Hawking, Bill Gates, Elon Musk, Henry Kissinger, et al. that superhuman AI is an existential threat to humanity? Such speculations are also wholly divorced from the present-day reality of AI technology and thus say very little about it. While conceptually fascinating as a thought experiment, these speculations distract from the more prosaic but far more likely risks posed by AI, as well as its many potential benefits as the technology matures.

In the high stakes military sphere, where a premium is placed on speed, autonomy, and deception, the temptation for strategic rivals to cut corners for short-term tactical gain is self-evident. After an initial surge of widespread speculation in the literature related to AI and security, this volume offers readers a sober discussion on the potential risks posed by AI to strategic stability between great military powers. The book demystifies the potential implications of 'military AI' and considers how and why it might become a fundamentally destabilizing technology in the second nuclear age. As this conversation unfolds, what becomes clear is that even if the trajectory of AI is less revolutionary, but instead incremental and mundane, its impact on the traditional assumption about nuclear deterrence, arms control, and crisis stability could be profound nonetheless.

I owe a great debt of thanks to my colleagues and friends at the James Martin Center for Nonproliferation Studies at Monterey, where many of the ideas for this book were developed and honed. A special thanks goes to Bill Potter for his ongoing support, enthusiasm, and unique insights, which have enriched the book. I would also like to acknowledge the encouragement, friendship, and support of my new colleagues in the School of Law and Government at Dublin City University, not least Iain McMenamin and John Doyle. The development of this book benefited tremendously from being part of the Policy on Nuclear Issues community, led by the Center for Strategic and International Studies in the US and the Royal United Services Institute in the UK. The book received financial support from the Faculty of Humanities and Social Sciences Book Publication Scheme at Dublin City University. My thanks must also go to the many experts who challenged my ideas and sharpened the arguments on the presentations I have given at international forums during the development of this book.

I am grateful for the generous support, mentorship, and deep knowledge of colleagues and friends, including: James Acton, John Amble, Zhang Baohui, John Borrie, Lyndon Burford, Jeffrey Cummings, Jeffrey Ding, Mona Dreicer, Sam Dunin, Andrew Futter, Erik Gartzke, Andrea Gilli, Rose Gottemoeller, Rebecca Hersman, Michael Horowitz, Patrick Howell, Alex Lennon, Keir Lieber, Jon Lindsay, Joe Maiolo, Giacomo Persi Paoli, Kenneth Payne, Tom Plant, Daryl Press, Adam Quinn, Andrew Reddy, Wyn Rees, Brad Roberts, Mick Ryan, Daniel Salisbury, John Shanahan, Wes Spain, Reuben Steff, Oliver Turner, Chris Twomey, Tristen Volpe, Tom Young, Benjamin Zala.

My thanks also to the excellent and professional team at Manchester University Press – especially editors Rob Byron and Jon de Peyer – for their support and encouragement, not to mention the anonymous reviewers, whose comments and suggestions kept my ego in check and arguments honest.

This book is dedicated to my adoring wife, Cindy. Without her love, patience, and kindness – not to mention technical savvy – the completion of this book would not have been possible. My amazing wife has given me unstinting encouragement – and a shoulder to cry on – in my metamorphosis from financier to academic.

# List of abbreviations

| | |
|---|---|
| AGI | artificial general intelligence |
| APT | advanced persistent threat |
| ASW | anti-submarine warfare |
| ATR | automatic target recognition |
| BRI | belt-and-road-initiative |
| C2 | command and control |
| C3I | command, control, communications, and intelligence |
| DARPA | Defense Advanced Research Projects Agency (US DoD) |
| DL | deep learning |
| DoD | Department of Defense (US) |
| GAN | generative adversarial network |
| HGV | hypersonic guide vehicle |
| ICBM | intercontinental ballistic missile |
| IoT | internet of things |
| IR | International Relations |
| ISR | intelligence, surveillance, and reconnaissance |
| JAIC | Joint Artificial Intelligence Center |
| LAMs | loitering attack munitions |
| LAWS | lethal autonomous weapon systems |
| MAD | mutually assured destruction |
| ML | machine learning |
| NC3 | nuclear command, control, and communications |
| NFU | no-first-use (nuclear pledge) |
| PLA | People's Liberation Army |
| PRC | People's Republic of China |
| R&D | research and development |
| SLBM | submarine launched ballistic missile |
| SSBN | nuclear-powered ballistic missile submarine |

| | |
|---|---|
| UAS | unmanned autonomous systems |
| UAVs | unmanned aerial vehicles |
| USVs | unmanned surface vehicles |
| UUVs | unmanned undersea vehicles |

# Introduction:
## opening the AI Pandora's box

The hype surrounding AI[1] has made it easy to overstate the opportunities and challenges posed by the development and deployment of AI in the military sphere.[2] Many of the risks posed by AI in the nuclear domain today are not necessarily new. That is, recent advances in AI (especially machine learning (ML) techniques) exacerbate existing risks to escalation and stability rather than generating entirely new ones. While AI could enable significant improvements in many military domains – including the nuclear enterprise – future developments in military AI will likely be far more prosaic than implied in popular culture.[3] The book's core thesis is deciphering, within a broad range of technologies, proven capabilities and applications from mere speculation.

After an initial surge in the literature related to AI and national security, broadly defined, more specificity in the debate is now required.[4] Whereas much ink has been spilled on the strategic implications of advanced technologies such as missile defense systems, anti-satellite weapons, hypersonic weapons, and cyberspace, the potential impact of the rapid diffusion and synthesis of AI on future warfare – in particular in a high-end strategic standoff between two dominant powers – has been only lightly researched.[5] The book addresses this gap and offers a sober assessment of the potential risks AI poses to strategic stability between great military powers. This assessment demystifies the hype surrounding AI in the context of nuclear weapons and, more broadly, future warfare. Specifically, it highlights the potential, multifaceted intersections of this disruptive technology with nuclear stability. The book argues that the inherently destabilizing effects of military AI may exacerbate tension between nuclear-armed great powers – especially China and the US – but not for the reasons you might think.

Since the mid-2010s, researchers have achieved significant milestones in the development of AI and AI-related technologies – either enabled or enhanced by AI or critical to the development of AI technology, inter alia, quantum technology and computing, big-data analytics,[6] the internet of things, miniaturization, 3D printing, tools for gene-editing, and robotics and autonomy. Moreover, these achievements occurred significantly faster than experts in the field anticipated.[7] For example, in 2014, the AI expert who designed the world's best Go-playing (or AlphaGo) program predicted that it would be another ten years before a computer could defeat a human Go champion.[8] Researchers at Google's DeepMind achieved this technological feat just one year later. The principal forces driving this technological transformation include: the exponential growth in computing performance; expanded data-sets;[9] advances in the implementation of ML techniques and algorithms (especially in the field of deep neural networks); and, above all, the rapid expansion of commercial interest and investment in AI – chapter 1 analyzes these forces.[10]

AI technologies could impact future warfare and international security in three interconnected ways:[11] amplifying the uncertainties and risks posed by *existing* threats (both in the physical and virtual domains); transforming the nature and characteristics of these threats; and introducing *new risks* to the security landscape. AI could portend fundamental changes to military power, in turn re-ordering the military balance of power and triggering a new military-technological arms race.[12] The potential threats posed by AI-augmented capabilities to nuclear security and stability considered in this book can be grouped under three broad categories:[13] *digital* (or cyber and non-kinetic) risks such as spear-phishing, speech synthesis, impersonation, automated hacking, and data poisoning (see chapters 8 and 9);[14] *physical* (or kinetic) risks such as hypersonic weapons, and drones in swarm attacks (see chapters 6 and 7); and *political* risks such as regime stability, political processes, surveillance, deception, psychological manipulation, and coercion – especially in the context of authoritarian states (see chapter 9).

World leaders have been quick to recognize the transformative potential of AI as a critical component of national security policy.[15] In 2017, Russian President Vladimir Putin asserted, "the one who becomes the leader in AI will be the ruler of the world."[16] As Part II of the book shows, while advances in AI technology are not necessarily a zero-sum game, first-mover advantages that rival states seek to achieve a monopolistic position will likely exacerbate strategic competition. In a quest to become a 'science and technology superpower,' and catalyzed by AlphaGo's victory (or China's 'Sputnik moment'), Beijing launched a national-level AI-innovation agenda for 'civil-military fusion' – or a US Defense Advanced Research Projects Agency with Chinese characteristics.[17] The Russian military has targeted 30 percent of its entire force structure to be robotic by 2025. In short, national-level objectives and initiatives demonstrate recognition by the global defense community of the transformative – or even military-technical revolutionary – potential of AI for states' national security and strategic objectives.[18]

Driven, in large part, by the perceived strategic challenge from rising revisionist and revanchist powers (notably China and Russia), the US Department of Defense (DoD) in 2016 released a 'National Artificial Intelligence Research and Development Strategic Plan' on the potential for AI to reinvigorate US military dominance.[19] According to then-US Deputy Secretary of Defense, Robert Work, "we cannot prove it, but we believe we are at an *inflection point in AI and autonomy*" (emphasis added).[20] The DoD also established the Defense Innovation Unit Experimental to foster closer collaboration between the Pentagon and Silicon Valley.[21] In sum, advances in AI could presage fundamental changes to military power, with implications for the re-ordering of the balance of power.[22]

Opinions surrounding the impact of AI on future warfare and international security range more broadly from *minimal* (technical and safety concerns within the defense community could lead to another 'AI Winter') to *evolutionary* (dramatic improvements in military effectiveness and combat potential, but AI innovations will unlikely advance beyond task-specific – or 'narrow' AI – applications that require human oversight), and, in extremis, *revolutionary* (a fundamental transformation of both the character and nature of warfare).[23] Some experts speculate that AI could push the pace of combat to a point where machine actions surpass the rate of human decision-making and potentially shift the cognitive bases of international conflict and warfare, challenging the Clausewitzian notion that war is a fundamentally human endeavor, auguring a genuine (and potentially unprecedented), revolution in military affairs (see chapter 3).[24] This book explores the tension between those who view AI's introduction into warfaring as inherently destabilizing and revolutionary, and juxtaposed, and those who view AI as more evolutionary and as a double-edged sword for strategic stability.

Former US Defense Secretary, James Mattis, warned that AI is "fundamentally different" in ways that question the very nature of war itself.[25] Similarly, the DoD's Joint Artificial Intelligence Center former Director, Lt. General Jack Shanahan, opined, "AI will change the character of warfare, which in turn will drive the need for wholesale changes to doctrine, concept development, and tactics, techniques, and procedures."[26] Although it is too early to speak of the development of AI-specific warfighting operational concepts (or 'Algorithm Warfare'), or even how particular AI applications will influence military power in future combat arenas,[27] defense analysts and industry experts are nonetheless predicting the potential impact AI might have on the future of warfare and the military balance.[28]

Notwithstanding its skeptics, a consensus has formed that the convergence of AI with other technologies will likely enable new capabilities and enhance existing advanced capabilities – offensive, defensive, and kinetic and non-kinetic – creating new challenges for national security. It is clear, however, that no single military power will dominate all of the manifestations of AI while denying its rivals these potential benefits. The intense competition in the development and deployment of military-use AI applications – to achieve the first-mover advantage in this disruptive technology – will exacerbate uncertainties about the future balance of military power, deterrence, and strategic stability, thereby, increasing risks of nuclear warfare.

Any strategic debate surrounding nascent, and highly classified, technologies such as AI comes with an important caveat. Since we have yet to see how AI might influence states' deterrence calculations and escalation management in the real world – and notwithstanding valuable insights from non-classified experimental wargaming – coupled with the uncertainties surrounding AI-powered capabilities (e.g. searching for nuclear-armed submarines), the discourse inevitably involves a measure of conjecture and inference from technical observations and strategic debate.[29]

## Core arguments

This book argues that military-use AI is fast becoming a principal potential source of instability and great-power strategic competition.[30] The future safety of military AI systems is not just a technical challenge, but also fundamentally a political and human one.[31] Towards this end, the book expounds on four interrelated core arguments.[32]

First, AI in isolation has few genuinely strategic effects. AI does not exist in a vacuum. Instead, it is a potential power force multiplier and enabler for advanced weapons (e.g. cyber capabilities, hypersonic vehicles, precision-guided missiles, robotics, and anti-submarine warfare), mutually reinforcing the destabilizing effects of these existing capabilities. Specifically, AI technology has not yet evolved to a point where it would allow nuclear-armed states to credibly threaten the survivability of each other's nuclear second-strike capability. For the foreseeable future, the development trajectory of AI and its critical-enabling technologies (e.g. 5G networks, quantum technology, robotics, big-data analytics, and sensor and power technology) means that AI's impact on strategic stability will likely be more prosaic and theoretical, than transformational.

Second, AI's impact on stability, deterrence, and escalation will be determined as much (or more so) by states' perception of its functionality than what it is – technically or operationally – capable of doing. Further, in addition to the importance of military force posture, capabilities, and doctrine, the effects of AI will continue to have a strong cognitive element (or human agency), which could increase the risk of inadvertent or accidental escalation caused by misperception or miscalculation.

Third, the concomitant pursuit of AI technology by great powers – especially China, the US, and Russia – will likely compound the destabilizing effects of AI in the context of nuclear weapons.[33] The concept of nuclear multipolarity (explored in chapter 4) is important precisely because different states will likely choose different answers to the new choices emerging in the digital age. Besides, an increasingly competitive and contested nuclear multipolar world order could mean that the potential military advantages offered by AI-augmented capabilities prove irresistible to states, in order to sustain or capture the technological upper hand over rivals.

Finally, against this inopportune geopolitical backdrop, coupled with the perceived strategic benefits of AI-enhanced weapons (especially AI and autonomy), the most pressing risk posed to nuclear security is, therefore, the early adoption of unsafe, unverified, and unreliable AI technology in the context of nuclear weapons, which could have catastrophic implications.[34]

The book's overarching goal is to elucidate some of the consequences of military-use AI's recent developments in strategic stability between nuclear-armed states and nuclear security. While the book is not a treatise on the technical facets of AI, it does not eschew the technical aspects of the discourse. A solid grasp of some of the key technological developments in the evolution of AI is a critical first step to determine what AI is (and is not), what it is capable of, and how it differs from other technologies (see chapter 1). Without a robust technical foundation, demystifying hype from reality, and pure speculation from informed inferences and extrapolation would be an insurmountable and ultimately fruitless endeavor.

## Book plan

This book is organized into three parts and eight chapters. Part I considers how and why AI might become a force for strategic instability in the post-Cold War system – or the second nuclear age. Chapter 1 defines and categorizes the current state of AI and AI-enabling technologies. It describes several possible implications of specific AI systems and applications in the military arena, in particular those that might impinge on the nuclear domain. What are the possible development paths and linkages between these technologies and specific capabilities (both existing and under development)? The chapter contextualizes the evolution of AI within the broader field of science and engineering in making intelligent machines. The chapter also highlights the centrality of ML and autonomous systems to understanding AI in the military sphere, at both an operational and strategic level of warfare. The purpose of the chapter is to demystify the military implications of AI and debunk some of the misrepresentations and hyperbole surrounding AI.

Chapter 2 presents the central theoretical framework of the book. By conceptualizing 'strategic stability' with the emerging technological challenges posed to nuclear security in the second nuclear age, the chapter tethers the book's core arguments into a robust analytical framework. It contextualizes AI within the broad spectrum of military technologies associated with the 'computer revolution.' The chapter describes the notion of 'military AI' as a natural manifestation of an established trend in emerging technology. Even if AI does not become the next revolution in military affairs, and its trajectory is more incremental and prosaic, the implications for the central pillars of nuclear deterrence could still be profound.

Part II turns to the strategic competition between China and the US. What is driving great military powers to pursue AI technologies? How might the proliferation and diffusion of AI impact the strategic balance? The section explains that as China and the US internalize these emerging technological trends, both sides will likely conceptualize them very differently. Scholarship on military innovation has demonstrated that – with the possible exception of nuclear weapons – technological innovation rarely causes the military balance to shift. Instead, *how and why* militaries employ a technology usually proves critical.[35] Part II also analyzes the increasingly intense rivalry playing out in AI and other critical technologies, and the potential implications of these developments for US–China crisis stability, arms races, escalation, and deterrence. How might the linkages between AI and other emerging technologies affect stability and deterrence?

Chapter 3 investigates the intensity of US–China strategic competition playing out within a broad range of AI and AI-enabling technologies. It considers how great-power competition is mounting in intensity within several dual-use high-tech fields, why these innovations are considered by the US to be strategically vital, and how (and to what end) the US responds to the perceived challenge posed by China to its technological hegemony. Why does the US view China's progress in dual-use emerging security technology as a threat to its first-mover advantage? How is the US responding to the perceived challenge to its military-technological leadership?

The chapter describes how great-power competition is mounting within several dual-use high-tech fields, why these innovations are considered by Washington to be strategically vital, and how (and to what end) the US responds to the perceived challenge posed by China to its defense innovation hegemony. The chapter uses the International Relations concept of 'polarity' to consider the shifting power dynamics in AI-related emerging security technologies. The literature on the diffusion of military technology demonstrates how states react to and assimilate defense innovations that can have profound implications for strategic stability and the likelihood of war.[36] It argues that the strategic competition playing out within a broad range of dual-use AI-enabling technologies will narrow the technological gap separating great military powers – notably China and the US – and, to a lesser extent, other technically advanced small–medium powers.

Chapter 4 considers the possible impact of AI-augmented technology for military escalation between great military powers, notably the US–China dyad. The chapter argues that divergent US–China thinking on the escalation (especially inadvertent) risks of co-mingling nuclear and non-nuclear capabilities will exacerbate the destabilizing effects caused by the fusion of these capabilities with AI applications. Under crisis conditions, incongruent strategic thinking – coupled with differences in regime type, nuclear doctrine, strategic culture, and force structure – might exacerbate deep-seated (and ongoing) US–China mutual mistrust, tension, misunderstandings, and misperceptions.[37]

The chapter demonstrates that the combination of first-strike vulnerability and opportunity afforded by advances in military technologies like AI will have destabilizing implications for military (especially inadvertent) escalation in future warfare. How concerned is Beijing about inadvertent escalation risks? Are nuclear and conventional deterrence or conventional warfighting viewed as separate categories by Chinese analysts? And how serious are the escalation risks arising from entanglement in a US–China crisis or conflict scenario?

Part III includes four case study chapters, which constitute the empirical core of the book. These studies consider the escalation risks associated with AI. They demonstrate how and why military AI systems fused with advanced strategic non-nuclear weapons (or conventional counterforce capabilities) could cause or exacerbate escalation risks in future warfare. They illuminate how these AI-augmented capabilities would work, and why, despite the risks associated with their deployment, great military powers will likely use them nonetheless.

Chapter 5 considers the implications of AI-augmented systems for the survivability and credibility of states' nuclear deterrence forces (especially nuclear-deterrent submarines and mobile missiles). How might AI-augmented systems impact the survivability and credibility of states' nuclear deterrence forces? The chapter argues that emerging technologies – AI, ML, and big-data analytics – will significantly improve the ability of militaries to locate, track, target, and destroy adversaries' nuclear-deterrent forces without the need to deploy nuclear weapons. Furthermore, AI applications that make hitherto survivable strategic forces more vulnerable (or are perceived as such) could have destabilizing escalatory effects. The chapter finds that specific AI applications (e.g. for locating mobile missile launchers) may be strategically destabilizing, not because they work too well but because they work just well enough to feed uncertainty.[38]

Chapter 6 examines the possible ways AI-augmented drone swarms and hypersonic weapons could present new challenges to missile defense, undermine states' nuclear-deterrent forces, and increase the escalation risks. The case study unpacks the possible strategic operations (both offensive and defensive) that AI-augmented drone swarms could enable, and the potential impact of these operations for crisis stability. It also considers how ML-enabled qualitative improvements to hypersonic delivery systems could amplify the escalatory risks associated with long-range precision munitions.

Chapter 7 elucidates how AI-infused cyber capabilities may be used to manipulate, subvert, or otherwise compromise states' nuclear assets. It examines the notion that enhanced cybersecurity for nuclear forces may simultaneously make cyber-dependent nuclear weapon systems more vulnerable to cyber-attacks. How could AI-augmented cyber capabilities create new pathways for escalation? How might AI affect the offense–defense balance in cyberspace? The chapter argues that future iterations of AI-powered cyber (offense and defense) capabilities will increase escalation risks. It finds that AI-enhanced cyber counterforce capabilities will further complicate the cyber-defense challenge, thereby increasing the escalatory effects of offensive cyber capabilities. Moreover, as the linkages between digital and physical systems expand, the ways an adversary might use cyber-attacks in both kinetic and non-kinetic attacks will increase.

Chapter 8, the final case study, considers the impact of military commanders using AI systems in the strategic decision-making process, despite the concerns of defense planners. In what ways will advances in military AI and networks affect the dependability and survivability of nuclear command, control, and communications systems? The study analyzes the risks and trade-offs of increasing the role of machines in the strategic decision-making process. The chapter argues that the distinction between the impact of AI at a tactical level and a strategic one is not binary. Through the discharge of its 'support role,' AI could, in critical ways, influence strategic decisions that involve nuclear weapons. Moreover, as emerging technologies like AI are superimposed on states' legacy nuclear support systems, new types of errors, distortions, and manipulations are more likely to occur – especially in the use of social media.

The concluding chapter reviews the book's core findings and arguments. The chapter finishes with a discussion on how states – especially great military powers – might mitigate, or at least manage, the escalatory risks posed by AI and bolster strategic stability as the technology matures. Implications and policy recommendations are divided into two closely correlated categories. First, enhancing debate and discussion; and second, specific policy recommendations and tools to guide policy-makers and defense planners as they recalibrate their national security priorities to meet the emerging challenges of an AI future. Of course, these recommendations are preliminary and will inevitably evolve as the technology itself matures.

## Notes

1 Artificial intelligence (AI) refers to computer systems capable of performing tasks requiring human intelligence, such as: visual perception, speech recognition, and decision-making. These systems have the potential to solve tasks requiring human-like perception, cognition, planning, learning, communication, or physical action – see chapter 1 for an AI military-use primer.
2 Speculations about superintelligent AI or the threat of superman AI to humanity – as chapter 1 explains – are entirely disconnected from anything factual about the capabilities of present-day AI technology. For example, see Mike Brown, "Stephen Hawking Fears AI May Replace Humans," *Inverse*, November 2, 2017, www.inverse.com/article/38054-stephen-hawking-ai-fears (accessed March 10, 2020); and George Dvorsky, "Henry Kissinger Warns That AI Will Fundamentally Alter Human Consciousness," *Gizmodo*, May 11, 2019, https://gizmodo.com/henry-kissinger-warns-that-ai-will-fundamentally-alter-1839642809 (accessed March 10, 2020).
3 For example, George Zarkadakis, *In Our Image: Savior or Destroyer? The History and Future of Artificial Intelligence* (New York: Pegasus Books, 2015); and Christianna Ready, "Kurzweil Claim That Singularity Will Happen By 2045," *Futurism*, October 5, 2017, https://futurism.com/kurzweil-claims-that-the-singularity-will-happen-by-2045 (accessed March 10, 2020).
4 In recent years, a growing number of IR studies have debated a range of issues relating to the 'AI question' – most notably legal, ethical, normative, economic, and technical aspects of the discourse. See for example, Max Tegmark, *Life 3.0: Being Human in the Age of Artifical Intelligence* (London: Penguin Random House, 2017); and Adam Segal, *Conquest in Cyberspace: National Security and Information Warfare* (Cambridge: Cambridge University Press, 2015). For a recent technical study on autonomous weapon systems, see Jeremy Straub, "Consideration of the Use of Autonomous, Non-Recallable Unmanned Vehicles and Programs as a Deterrent or Threat by State Actors and Others," *Technology in Society*, 44 (February 2016), pp. 1–112. For social and ethical implications see Patrick Lin, Keith Abney, and George Bekey (eds), *Robot Ethics: The Ethical and Social Implications of Robotics* (Cambridge, MA: MIT Press, 2014).
5 Notable exceptions include: Patricia Lewis and Unal Beyza, *Cybersecurity of Nuclear Weapons Systems: Threats, Vulnerabilities and Consequences* (London: Chatham House Report, Royal Institute of International Affairs, 2018); Mary L. Cummings, *Artificial Intelligence and the Future of Warfare* (London: Chatham

House, 2017); Lawrence Freedman, *The Future of War* (London: Penguin Random House, 2017); Lucas Kello, *The Virtual Weapon and International Order* (New Haven: Yale University Press, 2017); Pavel Sharikov, "Artificial Intelligence, Cyberattack, and Nuclear Weapons – A Dangerous Combination," *Bulletin of the Atomic Scientists*, 74 6 (2018), pp. 368–373; Kareem Ayoub and Kenneth Payne, "Strategy in the Age of Artificial Intelligence," *Journal of Strategic Studies* 39, 5–6 (2016), pp. 793–819; and James S. Johnson, "Artificial Intelligence: A Threat to Strategic Stability," *Strategic Studies Quarterly*, 14, 1 (2020), pp. 16–39.

6 I. Emmanuel and C. Stanier, "Defining big data," in *Proceedings of the International Conference on Big Data and Advanced Wireless Technologies* (New York: ACM, 2016).

7 Recent progress in AI falls within two distinct fields: (1) 'narrow' AI and, specifically, machine learning; and (2) 'general' AI, which refers to AI with the scale and fluidity akin to the human brain. 'Narrow' AI is already in extensive use for civilian tasks. Chapter 1 will explain what AI is (and is not), and its limitations in a military context.

8 'Go' is a board game, popular in Asia, with an exponentially greater mathematical and strategic depth than chess.

9 'Machine learning' is a concept that encompasses a wide variety of techniques designed to identify patterns in, and learn and make predictions from, data-sets (see chapter 1).

10 Greg Allen and Taniel Chan, *Artificial Intelligence and National Security* (Cambridge, MA: Belfer Center for Science and International Affairs, 2017).

11 James Johnson, "Artificial Intelligence & Future Warfare: Implications for International Security," *Defense & Security Analysis*, 35, 2 (2019), pp. 147–169.

12 James Johnson, "The End of Military-Techno *Pax Americana*? Washington's Strategic Responses to Chinese AI-Enabled Military Technology," *The Pacific Review*, www.tandfonline.com/doi/abs/10.1080/09512748.2019.1676299?journalCode=rpre20 (accessed February 5 2021).

13 Center for a New American Security, University of Oxford, University of Cambridge, Future of Humanity Institute, OpenAI & Future of Humanity Institute, *The Malicious Use of Artificial Intelligence: Forecasting, Prevention, and Mitigation* (Oxford: Oxford University, February 2018) https://arxiv.org/pdf/1802.07228.pdf (accessed March 10, 2020).

14 These AI vulnerabilities are, however, distinct from traditional software vulnerabilities (e.g. buffer overflows). Further, they demonstrate that even if AI systems exceed human performance, they often fail in unpredictable ways that a human never would.

15 Robert O. Work, *Remarks by Defense Deputy Secretary Robert Work at the CNAS Inaugural National Security Forum* (Washington, DC: CNAS, July 2015), www.defense.gov/Newsroom/Speeches/Speech/Article/634214/cnas-defense-forum/ (accessed March 10, 2020).

16 James Vincent, "Putin Says the Nation that Leads in AI 'Will be the Ruler of the World,'" *The Verge*, September 4, 2017, www.theverge.com/2017/9/4/16251226/russia-ai-putin-rule-the-world (accessed March 10, 2020).

17 The State Council Information Office of the People's Republic of China, "State Council Notice on the Issuance of the New Generation AI Development Plan," July 20, 2017, www.gov.cn/zhengce/content/2017-07/20/content_5211996.htm (accessed March 10, 2020).
18 A military-technical revolution has is associated with periods of sharp, discontinuous change that make redundant or subordinate existing military regimes, or the most common means for conducting war (see chapter 2).
19 National Science and Technology Council, *The National Artificial Intelligence Research and Development Strategic Plan* (Executive Office of the President of the US, Washington, DC, October 2016), www.nitrd.gov/PUBS/national_ai_rd_strategic_plan.pdf.
20 *Reagan Defense Forum: The Third Offset Strategy* (Washington, DC, US Department of Defense. November 7, 2015), https://dod.defense.gov/News/Speeches/Speech-View/Article/628246/reagan-defense-forum-the-third-offset-strategy/ (accessed March 10, 2020). Recent defense initiatives that have applied deep-learning techniques to autonomous systems include: the US Air Force Research Laboratory's Autonomous Defensive Cyber Operations; National Geospatial Agency's Coherence Out of Chaos program (deep-learning-based queuing of satellite data for human analysts); and Israel's Iron Dome air defiance system.
21 Fred Kaplan, "The Pentagon's Innovation Experiment," *MIT Technology Review*, December 16, 2016, www.technologyreview.com/s/603084/the-pentagons-innovation-experiment/ (accessed March 10, 2020).
22 In addition to AI, China and Russia have also developed other technologically advanced (and potentially disruptive) weapons such as: cyber warfare tools; stealth and counter-stealth technologies; counter-space, missile defense, and guided precision munitions (see chapters 4 and 5). See Timothy M. Bonds, Joel B. Predd, Timothy R. Heath, Michael S. Chase, Michael Johnson, Michael J. Lostumbo, James Bonomo, Muharrem Mane, and Paul S. Steinberg, *What Role Can Land-Based, Multi-Domain Anti-Access/Area Denial Forces Play in Deterring or Defeating Aggression?* (Santa Monica, CA: RAND Corporation, 2017), www.rand.org/pubs/research_reports/RR1820.html (accessed March 10, 2020).
23 Experts have raised concerns that AI could cause humans to lose control of military escalation management (i.e. the ability to influence the outcome or terminate conflict promptly), which could worsen strategic instability. See Jurgen Altmann and Frank Sauer, "Autonomous weapons and strategic stability," *Survival*, 59, 5, (2017), pp. 121–127; and Michael O'Hanlon, *Technological Change and the Future of Warfare* (Washington, DC: Brookings Institution, 2011).
24 For example, China's military leadership recently asserted that AI would lead to a profound military revolution. Elsa B. Kania, *Battlefield Singularity: Artificial Intelligence, Military Revolution, and China's Future Military Power* (Washington, DC: Center for a New American Security, 2017), p. 8.
25 "Press Gaggle by Secretary Mattis En Route to Washington, D.C.," Department of Defense, February 17, 2018, www.defense.gov/News/Transcripts/Trans

cript-View/Article/1444921/press-gaggle-by-secretary-mattis-en-route-to-wash ington-dc/ (accessed March 10, 2020).

26 Statement by Lt. General John Shanahan, then Director of the DoD JAIC, *Before the Senate Armed Services Committee on Emerging Threats and Capabilities on 'AI Initiatives,'* March 12, 2019, www.armed-services.senate.gov/imo/media/doc/Shanahan_03-12-19.pdf (accessed March 10, 2020).

27 Industry experts remain in disagreement about the potential trajectory of AI technology. To date, there is no consensus on the precise definition of central concepts, including autonomy, automation, or even AI itself (see chapter 1).

28 The historical record demonstrates that previous technological revolutions are evident only in hindsight. Thus, the influence of AI may not be apparent until it is deployed in a combat scenario. MacGregor Knox and Williamson Murray, "The future behind us," in *The Dynamics of Military Revolution, 1300–2050* (Cambridge: Cambridge University Press, 2001), p. 178.

29 Reid B. and C. Pauly, "Would US Leaders Push the Button? Wargames and the Sources of Nuclear Restraint," *International Security*, 43, 2 (2018), pp. 151–192; Jacquelyn G. Schneider, "Cyber Attacks on Critical Infrastructure: Insights From War Gaming," *War on the Rocks*, July 26, 2017, https://warontherocks.com/2017/07/cyber-attacks-on-critical-infrastructure-insights-from-war-gaming/ (accessed 8 January 2021); and Erik Lin-Greenberg, "Wargame of Drones: Remotely Piloted Aircraft and Crisis Escalation," August 22, 2020, *SSRN*, https://ssrn.com/abstract=3288988 (accessed March 10, 2020).

30 At the heart of strategic stability is limiting the incentives for states to launch a first nuclear strike, and thus reducing the conditions under which countries face pressures to escalate a conflict. Non-nuclear technologies with strategic effects (such as AI) can adversely disrupt these risks (see chapter 2).

31 The historical record demonstrates that the outbreak of war is often better explained by politics, rather than by changes in the military balance caused by emerging technology. See Keir A. Lieber, *War and the Engineers: The Primacy of Politics over Technology* (Ithaca, NY: Cornell University Press, 2005).

32 Johnson, "Artificial Intelligence & Future Warfare: Implications for International Security," pp. 147–169.

33 While the book's primary focus is the impact of AI for strategic stability between great military powers (especially China and the US), it also considers how the diffusion of AI technology to non-state actors (terrorists, criminals, and proxy-state actors), and non-nuclear states, might impact the strategic environment, thus, making nuclear warfare more or less likely. Chapters 4 and 9 examine these issues directly, while other sections of the book do so more tangentially.

34 The DoD has long recognized these kinds of concerns. In 1991, the Air Force's Tacit Rainbow anti-radiation missile program, which incorporated elements of Unmanned Aerial Vehicles and cruise missiles, was canceled, in large part because of the risk of error posed by autonomous systems used in offensive missions.

35 See Barry R. Posen, *The Sources of Military Doctrine: France, Britain, and Germany between the World Wars* (Ithaca, NY: Cornell Studies in Security Affairs, 1986).

36 See Michael C. Horowitz, *The Diffusion of Military Power: Causes and Consequences for International Politics* (Princeton, NJ: Princeton University Press, 2010); Gregory D. Koblentz, *Council Special Report. Strategic Stability in the Second Nuclear Age* (New York: Council on Foreign Relations Press, 2014).
37 James Johnson, "Washington's Perceptions and Misperceptions of Beijing's Anti-Access Area-Denial (A2–AD) 'Strategy': Implications for Military Escalation Control and Strategic Stability," *The Pacific Review*, 30, 3 (2017), pp. 271–288.
38 Advances in AI-augmented ISR systems could mitigate some of the uncertainties associated with tracking and targeting mobile nuclear missile launchers and make them more vulnerable to pre-emptive attacks. Edward Geist and Andrew Lohn, *How Might Artificial Intelligence Affect the Risk of Nuclear War?* (Santa Monica, CA: RAND Corporation, 2018), www.rand.org/content/dam/rand/pubs/perspectives/PE200/PE296/RAND_PE296.pdf (accessed March 10, 2020), p. 15.

# Part I

Destabilizing the AI renaissance

# 1

# Military AI primer

What is AI, and how does it differ from other technologies? What are the possible development paths and linkages between these technologies and specific capabilities, both existing and under development? This chapter defines and categorizes the current state of AI and AI-enabling technologies.[1] The chapter highlights the centrality of machine learning (ML),[2] and autonomous systems (or 'machine autonomy'),[3] to understanding AI in the military sphere and the potential uses of these nuanced approaches in conjunction with AI at both an operational and strategic level of warfare.

The chapter proceeds in two sections. The first contextualizes the evolution of AI within the field of science and engineering to make intelligent machines. It defines AI as a universal term that can improve automated systems' performance to solve a wide variety of complex tasks. Next, it describes some of AI's limitations in order to clearly understand what AI is (and what it is not), and how best to implement AI in a military context. This section ends with a brief primer on ML's critical role as an enabler (and subset) of AI, based on computational systems that can learn and teach through a variety of techniques.

The second section demystifies the military implications of AI and critical AI-enabling technology. It debunks some of the misrepresentations and hyperbole surrounding AI. Then it describes how ML and autonomy could intersect with nuclear security in a multitude of ways, with both positive and negative implications for strategic stability – chapters 2 and 8 return to this idea. Next, it conceptualizes AI-augmented applications into those that have predominately operational, tactical, and strategic consequences in future warfare. This section argues that while the potential tactical and operational impact of AI is qualitatively self-evident, its effect at a strategic level remains uncertain.

## What is (and is not) AI?

AI research began as early as the 1950s as a broad concept concerned with the science and engineering of making intelligent machines.[4] In the decades that followed, AI research went through several development phases – from early exploitations in the 1950s and 1960s, the 'AI Summer' during the 1970s through to the early 1980s, and to the 'AI Winter' from the 1980s. Each of these phases failed to live up to its initial, and often over-hyped, expectations – in particular, when *intelligence* has been confused with *utility*.[5] Since the early 2010s, the explosion of interest in the field (or the 'AI renaissance') occurred due to the convergence of four critical enabling developments:[6] the exponential growth in computing processing power and cloud computing; expanded data-sets (especially 'big-data' sources);[7] advances in the implementation of ML techniques and algorithms (especially deep 'neural networks');[8] and the rapid expansion of commercial interest and investment in AI technology.[9]

AI is concerned with machines that emulate capabilities which are usually associated with human intelligence, such as language, reasoning, learning, heuristics, and observation. Today, all practical (i.e. technically feasible) AI applications fall into the 'narrow' category, and less so, artificial general intelligence (AGI) – or 'superintelligence.' Narrow AI has been broadly used in a wide range of civilian and military tasks since the 1960s,[10] and involves statistical algorithms (mostly based on ML techniques) that learn procedures through analysis of large training data-sets designed to approximate and replicate human cognitive tasks.[11] 'Narrow AI' is the category of AI that this book refers to when it assesses the impact of this technology in a military context.

Most experts agree that the development of AGI is at least several decades away, if feasible at all.[12] While the potential of AGI research is high, the anticipated exponential gains in the ability of AI systems to provide solutions to problems today are limited in scope. Moreover, these narrow-purpose applications do not necessarily translate well to more complex, holistic, and open-ended environments (i.e. modern battlefields), which exist simultaneously in the virtual (cyber/non-kinetic) and physical (or kinetic) plains.[13]

That is not to say, however, the conversation on AGI, and its potential impact should be entirely eschewed. If and when AGI does emerge, then ethical, legal, and normative frameworks will need to be devised to anticipate the implications for what would be a potentially pivotal moment in the course of human history. To complicate matters further, the distinction between narrow and general AI might prove less of an absolute (or binary) measure. Thus, research on narrow AI applications, such as game playing, medical diagnosis, and travel logistics, often results in incremental progress on general-purpose AI – moving researchers closer to AGI.[14]

AI has generally been viewed as a sub-field of computer science, focused on solving computationally hard problems through search, heuristics, and probability. More broadly, AI also draws heavily from mathematics, human psychology and biology, philosophy, linguistics, psychology, and neuroscience (see Figure 1.1).[15] Because of the divergent risks involved and development timeframes in the two distinct types of AI, the discussion in this book is careful not to conflate them.[16] Given the diverse approaches to research in AI, there is no universally accepted definition of AI[17] – confusing when the generic term 'Artificial Intelligence' is used to make grandiose claims about its revolutionary impact on military affairs, or 'revolution in military affairs.'[18] Moreover, if AI is defined too narrowly or too broadly, we risk understating the potential scope of AI capabilities; or, juxtaposed,

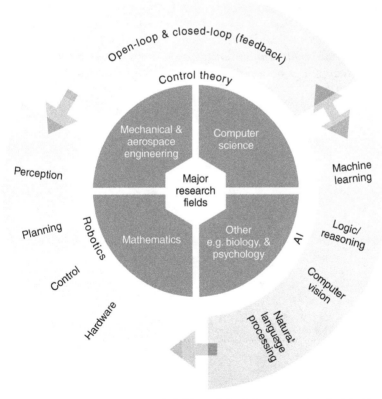

**Figure 1.1** Major research fields and disciplines associated with AI

fail to specify the unique capacity that AI-powered applications might have, respectively. A recent US congressional report defines AI as follows:

> Any artificial system that performs tasks under varying and unpredictable circumstances, *without significant human oversight*, or that can learn from their experience and improve their performance ... they may solve tasks requiring human-like perception, cognition, planning, learning, communication, or physical action (emphasis added).[19]

In a similar vein, the US DoD recently defined AI as:

> The ability of machines to *perform tasks that normally require human intelligence* – for example, recognizing patterns, learning from experience, drawing conclusions, making predictions, or taking action – whether digitally or as the smart software behind autonomous physical systems (emphasis added).[20]

AI can be best understood as a universal term for improving the performance of automated systems to solve a wide variety of complex tasks including:[21] *perception* (sensors, computer vision, audio, and image processing); *reasoning and decision-making* (problem-solving, searching, planning, and reasoning); *learning and knowledge representation* (ML, deep networks, and modeling); *communication* (language processing); *automatic* (or autonomous) systems and robotics (see Figure 1.2); and *human–AI collaboration*

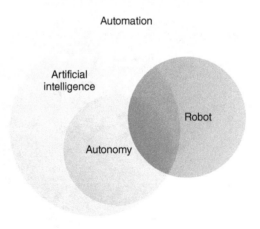

**Figure 1.2** The linkages between AI & autonomy

(humans define the systems' purpose, goals, and context).[22] As a potential enabler and force multiplier of a portfolio of capabilities, therefore, military AI is more akin to electricity, radios, radar, and intelligence, surveillance, and reconnaissance (ISR) support systems than a 'weapon' per se.[23]

AI suffers from several technical shortcomings that should prompt prudence and restraint in the early implementation in a military context. Today, AI systems are brittle and can only function within narrowly pre-defined problem-sets and context parameters.[24] Specifically, AI cannot effectively and reliably diagnose errors (e.g. sampling errors or intentional manipulation) from complex data-sets, and the esoteric mathematics underlying AI algorithms.[25] Further, AI systems are unable to handle novel situations reliably; AI relies on a posteriori knowledge to make inferences and inform decision-making. Failure to execute a particular task, especially if bias results are generated, would likely diminish the level of trust placed in these applications.[26] Therefore, in order to mitigate the potentially destabilizing effects from either poorly conceptualized an accidental-prone AI – or states' exaggerating (or underestimating) the strategic impact of military AI capabilities[27] – decision-makers must better understand what AI is (and what it is not), its limitations, and how best to implement AI in a military context.[28]

## Machine learning: a critical AI-enabler but no 'alchemy'[29]

ML is an approach to software engineering developed during the 1980s and 1990s, based on computational systems that can 'learn'[30] and teach themselves through a variety of techniques, such as neural networks, memory-based learning, case-based reasoning, decision-trees, supervised learning, reinforcement learning, unsupervised learning, and, more recently, generative adversarial networks. Consequently, the need for cumbersome human hand-coded programming has been dramatically reduced.[31]

From the fringes of AI until the 1990s, advances in ML algorithms with more sophisticated connections (i.e. statistics and control engineering) emerged as one of the most prominent AI methods (see Figure 1.3). In recent years, a subset of ML, deep learning (DL), has become the avant-garde AI software engineering approach, transforming raw data into abstract representations for a range of complex tasks, such as image recognition, sensor data, and simulated interactions (e.g. game playing).[32] The strength of DL is its ability to build complex concepts from simpler representations.[33]

Alongside the development of AI and ML, a new ecosystem of AI sub-fields and enablers have evolved, including: image recognition, machine vision, predictive analysis and planning, reasoning and representation, natural language representation and processing, robotics, and data

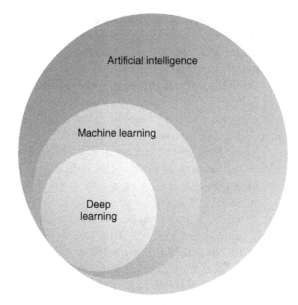

**Figure 1.3** Hierarchical relationship: ML is a subset of AI, and DL is a subset of ML

classification (see Figure 1.4).[34] In combination, these techniques have the potential to enable a broad spectrum of increasingly autonomous applications, inter alia: big-data mining and analytics; AI voice assistants; language and voice recognition aids; structured query language data-basing; autonomous weapons and autonomous vehicles; and information gathering and analysis, to name a few. One of the critical advantages of ML is that human engineers no longer need to explicitly define the problem to be resolved in a particular operating environment.[35] ML image recognition systems can be used, for example, to express mathematically the differences between images, which human hard-coders struggle to do.

ML's recent success can be attributed in large part to the rapid increase in computing power and the availability of vast data-sets to train ML algorithms. Today, AI ML techniques are routinely used in many everyday applications, including: empowering navigation maps for ridesharing software; by banks to detect fraudulent and suspicious transactions; making recommendations to customers on shopping and entertainment websites; supporting virtual personal assistants that use voice recognition software to offer their users content, and enabling improvements in medical diagnosis and scans.[36]

While advances in ML-enabled AI applications for identification, reconnaissance, and surveillance performance – used in conjunction with human

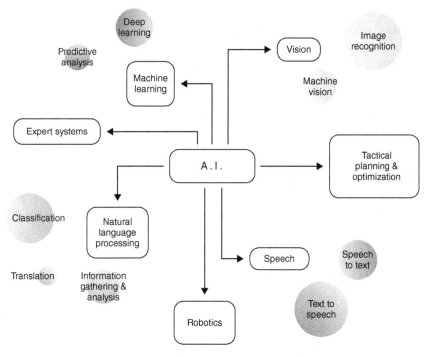

Figure 1.4 The emerging 'AI ecosystem'

intelligence analysis – could improve the ability of militaries to locate, track, and target an adversary's nuclear assets (see chapter 5),[37] three major technical bottlenecks remain unresolved: a dearth of quality data, automated detection weaknesses, and the so-called 'black box' (or explainability) problem-set.

### A dearth of quality data

ML systems are dependent on vast amounts of high-quality prelabeled data-sets – with both positive and negative examples – to learn and train from, in particular to reduce the risk of human bias and to certify methods of testing and verification. That is, ML AI is only as good as the data and information it was trained on and supplied with during operations. No matter how sophisticated the data-sets are, however, they are unable to replicate real-world situations perfectly. Each particular situation includes some irreducible error (or an error due to variance) because of incomplete and imprecise measurements and estimates. According to the former US DoD's Joint Artificial Intelligence Center Director Lt. General Jack Shanahan, "if you train [ML systems] against a very clean, gold-standard data-set, it will *not work in real-world conditions*" (emphasis added).[38] Further, relatively easy and cheap efforts (or 'adversarial AI') to fool these systems would likely render even the most sophisticated systems redundant (chapter 7 will consider the nature and impact of 'adversarial AI.')[39]

As a corollary, an ML system that memorizes data within a particular training set may fail when it is exposed to new and unfamiliar data. Because there is an infinite number of engineering options, it is impossible, therefore, to tailor an algorithm for all eventualities. Thus, as new data emerges, a near-perfect algorithm will likely quickly become redundant. In a military context, there is minimal room for error in balancing the pros and cons of deciding how much data to supply ML systems to learn from – or the so-called 'bias-variance' trade-off.[40] Even if a method was developed that performs flawlessly on the data fed to it, there is no guarantee that the system will perform in the same way on images it receives subsequently.[41]

In other words, there is no way to know for sure how ML-infused autonomous weapon systems such as AI-enhanced conventional counterforce capabilities (see chapters 5–7), would function in the field. Absent of the critical feedback from testing, validation, prototyping, and live testing (associated with the development of kinetic weapon systems), AI-enhanced capabilities will be prone to errors and accidents.[42] Because there is a naturally recursive relationship in how AI systems interact with humans and information, the likelihood of errors may be both amplified and reduced (chapter 8 will examine the nature and implications of this interaction).

Three technical limitations contribute to this data shortage problem.[43] First, in the military sphere, AI systems have relatively few images to train on (e.g. mobile road and rail-based missile launchers). This data imbalance will cause an AI system to maximize its accuracy by substituting images where it has a greater abundance of data, which may result in false positives.[44] In other words, in order to maximize its accuracy in classifying images, ML algorithms are incentivized to produce false negatives, such as misclassifying a regular truck as a mobile missile launcher. Thus, without the ability to distinguish between true and false, *all moving targets* could be considered viable targets by AI systems.

Second, AI is at a very early stage of 'concept learning.'[45] Images generally poorly depict reality. Whereas humans can deduce – using common sense – the function of an object from its external characteristics, AI struggles with this seemingly simple task. In situations where an object's form explicitly tells us about its function (i.e. language processing, speech, and handwriting recognition), this is less of an issue, and narrow AI generally performs well. However, in situations where an object's appearance does not offer this kind of information, AI's ability to induce or infer is limited. Thus, the ability of ML algorithms to reason is far inferior to human conjecture and criticism.[46] For example, in a military context, AI would struggle to differentiate the military function of a vehicle or platform. This problem is compounded by the shortage of quality data-sets, and the likelihood of AI adversarial efforts poised to exploit this shortcoming.

Third, ML becomes exponentially more difficult as the number of features, pixels, and dimensionality increases.[47] Greater levels of resolution and dimensional complexity – requiring more memory and time for ML algorithms to learn – could mean that images begin increasingly difficult for AI to differentiate. For example, a rail-based mobile missile launcher might appear to AI as a cargo train car or a military aircraft as a commercial airliner. In short, similar objects will become increasingly dissimilar to the AI, while images of different and unrelated objects will become increasingly indistinguishable.[48]

## Automated image-detection limitations

ML's ability to autonomously detect and cue precision-guided munitions is limited, particularly in cluttered and complex operational environments. This automated image recognition and detection weakness is caused largely by the poorly understood AI systems' ability to mimic human vision and cognition, which is notoriously opaque and irrational.[49] This limitation could, for example, reduce a commander's confidence in the effectiveness of AI-augmented counterforce capabilities, with mixed implications for strategic stability. Further, strategic competitors will continue to pursue countermeasures (e.g. camouflage, deception, decoys, and concealment) to protect their strategic forces against these advances.

## Black box problem-set

ML algorithms today are inherently opaque – especially those built on neural networks – creating 'black box' computational mechanisms that human engineers frequently struggle to fathom.[50] This 'black box' problem-set may cause unpredictability, for instance algorithms reacting in unexpected ways to data-sets used during the training phase, which could have severe consequences at a strategic level – not least, complicating the problem of attributing responsibility and accountability in the event of an accident or error in the safety-critical nuclear domain. Taken together, a dearth of quality data-sets for ML algorithms to learn from, automated detection technical limitations, the likelihood of adversarial countermeasures and exploitation, and the opacity of ML algorithms will significantly reduce the a priori knowledge AI systems can obtain from a situation – especially complex adversarial environments.

## Demystifying the military implications of AI

The convergence of several recent technological advances associated with the so-called 'Fourth Industrial Revolution,' – including quantum computing, ML (and its subset DL and neural networks), big-data analytics, robotics, additive manufacturing, nanotechnology, biotechnology, digital fabrication, and AI – will likely be leveraged by militaries to sustain or capture technological superiority on the future battlefield vis-à-vis strategic competitors (see chapter 3).[51] As described in the opening chapter, the implications of the convergence of these potentially transformative developments for strategic stability and nuclear security have been lightly researched.[52] This book will address some of the gaps in the literature; in particular, it demystifies the potential military implications of AI for nuclear risk, strategic stability, deterrence, and escalation management between rival nuclear-armed actors.

Today, a significant gap exists between the reality of what AI technology is capable of doing in a military context, and the expectations and fears of public opinion, policy-makers, and global defense communities. The misrepresentations that exist today in the narratives surrounding AI (especially in societal, economic, and national security debates) are, in large part, caused by the hyperbole and exaggerated depictions of AI in popular culture, and especially science fiction.[53] To be sure, frequent misrepresentations of the potential opportunities and risks associated with this technology in the military sphere can obscure constructive and sober debate on these topics. Specifically, this includes the trade-off between the potential operational, tactical, and strategic benefits of leveraging military AI, while effectively managing and mitigating the risks posed to nuclear security in the pursuit of these perceived first-mover advantages (see chapter 3).

In order to design dependable autonomous weapon systems, the following issues must first be addressed:[54] (1) improvements to machine observation and interpretation; (2) managing autonomous systems in the context of uncertainty, complexity, and incomplete information; (3) designing decision-making support applications that can enable ML algorithms to reliably follow (i.e. not deviate from) pre-set goals as intended by its human operator; (4) effectively balancing safety and predictability with the ability to operate at machine speed to win wars; (5) managing the integration of human and machine systems to optimize the exploitation of all the available data. Cognizant of these kinds of issues, US Strategic Operations Command (SOCOM) recently published a roadmap that invests heavily in AI and ML technology based on three overarching goals to create an algorithmic warfare multi-operational team: 'AI-ready workforce, AI-applications, and AI-outreach.'[55]

Recent advances in ML and AI have already led to significant qualitative improvements to a broad range of autonomous weapon systems, resolving several technical bottlenecks in existing military technology:

- Enhancing the detection capabilities of existing (nuclear and non-nuclear) early warning systems.
- Improving the collection and cross-analysis of ISR information to reduce the risk of miscalculation and inadvertent and accidental escalation.
- Bolstering the cyber defenses of nuclear and conventional command and control networks.
- Improving the allocation, procurement, and management of military resources.
- Creating new and innovative possibilities for arms control, testing, verification, and monitoring nuclear stockpiles – without the need for actually testing nuclear weapons.
- Enhancing conventional counterforce capabilities (e.g. cyber, electronic jamming, anti-satellite and hypersonic weapons, and missile defense systems).
- Enabling the deployment of unmanned autonomous systems – especially drones – in complex missions, and in hitherto inaccessible and cluttered environments (e.g. under-sea anti-submarine warfare).[56]

The multitude of ways AI and autonomy could intersect with nuclear security could have both positive and negative implications for strategic stability – Part III will consider the AI stability vs. stability-detracting debate.

Conceptually speaking, AI-augmented applications can be categorized into those with predominately operational, tactical, and strategic effects on warfare. At the operational and tactical level, applications include: autonomy and robotics (especially 'drone swarming,' see Figure 1.2); multi-actor interaction red-teaming and wargaming;[57] big data-driven modeling;[58] intelligence collection and analysis (e.g. to locate and track mobile missiles and troops movement);[59] communications resilience and cybersecurity; recruitment, training, and talent management; predictive maintenance, logistics, planning, and forecasting; and vendor contract and budget management.

Bayesian AI techniques that can detect deviations and optimize routes under variable conditions could enhance military logistics planning methods.[60] Similarly, as discussed in chapter 7, cyber-defense tools (e.g. network mapping and vulnerability identification and patching) for recognizing changes to patterns of behavior in a network and detecting anomalies, could benefit from automated AI rule-based reasoning approaches.[61] For example, Defense Advanced Research Projects Agency's

2016 Cyber Grand Challenge demonstrated the potential force multiplier effects of AI-cyber innovations.[62]

At a strategic level, warfare uses include: qualitative improvements to the nuclear ISR, and the nuclear command, control, and communications (NC3) networks; enhancing target acquisition, tracking, guidance systems, and discrimination of missile and air defense systems; force multipliers of *both* offensive and defensive ML-infused cyber capabilities; and qualitatively bolstering nuclear and non-nuclear missile delivery systems – including hypersonic variants.[63] For example, AI ML algorithms, together with advances in sensor technology, might allow nuclear delivery systems to operate autonomously, equipped with more robust countermeasures against jamming or spoofing attacks (chapter 7 will return to this theme).

While the potential tactical and operational impact of AI is today qualitatively axiomatic, its effect at a strategic level – especially in assessments of military power and strategic intention – remains uncertain. On the one hand, future AI-augmented command and control (C2) systems might overcome many of the shortcomings inherent to human strategic decision-making during wartime, such as a susceptibility to invest in sunk costs, skewed risk judgment, heuristics, and group-think – or the 'fog and friction of war.'[64] On the other hand, AI systems that enable commanders to predict the potential production, commissioning, deployment, and use of nuclear weapons by adversaries will likely lead to unpredictable system behavior and subsequent outcomes, which in extremis could undermine first-strike stability – the premise of mutually assured destruction – and make future nuclear war winnable. How might variations such as regime type, nuclear doctrine, strategy, strategic culture, or force structure make states more or less predisposed to developing AI in the nuclear domain? Chapter 8 will consider these issues.[65]

Despite the speed, diverse data pools, and processing power of algorithms compared to humans, these complex systems will still be dependent on the assumptions encoded into them by human engineers in order to simplify and extrapolate – potentially erroneous or bias – inferences from complexity, which result in unintended outcomes.[66] Specifically, biases (implicit or explicit) baked into image or text identification-sensing algorithms could cause errors from feedback loops – especially in cluttered and complex battlefield environments. For example, predictive policing AI systems that are fed with biased data can cause over-policing of marginalized groups in society, because those communities are over-represented in police data.[67] Moreover, different cultural biases intrinsic to particular ML training data-sets might also increase these risks of machine-to-machine interactions.[68]

In short, AI military systems – because of their ML training data-sets and learning trajectories – are cultural objects in their own right. Under the conditions of friction and the 'fog of war' – the inevitable uncertainties, misinformation, or even breakdown of organized units, which influence warfare – will influence their behavior, blind spots, and in-attentional blindness, allowing for errors and exploitation.[69] How, and by whom, should data be collected, stored, and processed? Who should be responsible for writing algorithms, and grounded to what kind of regulations, norms, ethics, or laws? In the words of military strategist Colin Gray, "the map of fog and friction is a living, dynamic one that reorganizes itself to frustrate the intrepid explorer."[70]

Besides, the added complexity of AI systems will likely amplify existing uncertainties about the value, scope, availability, credibility, and interpretation of information.[71] For the foreseeable future, therefore, narrow AI-infused sensing, self-learning, intelligence gathering, and analysis, and decision-making support systems, will continue to exhibit a similar penchant for cognitive bias and subjectively (e.g. attribution error, decision-making heuristics, path-dependency, and cognitive dissonance) that has long plagued the 'human' foreign policy and national security decision-making process.[72]

Notwithstanding major scientific breakthroughs in AI, incremental steps along the path already established would be sufficient to realize several technical goals that have long-eluded defense planners. For example, recent improvements in AI ML techniques will likely improve the capabilities of multiple sub-systems that rely on data-processing and pattern recognition capabilities, such as automatic target recognition (ATR), autonomous sensor platforms, vision-based guidance systems, malware detection software, and anti-jamming systems.[73] The Pentagon's Project Maven, for example, has already demonstrated the benefits of AI ML algorithms and image recognition to hunt for North Korean and Russian mobile missile launchers, and support aerial surveillance counter-terrorism operations against the Islamic State of Iraq and the Levant (ISIL).[74]

Moreover, advances in AI have the potential to significantly improve machine vision and other signal processing applications, which may overcome the main technical barriers for tracking and targeting adversaries' nuclear forces such as sensing, image processing, and estimating weapon velocities and kill radius (see chapter 5).[75] In combination, DL techniques, image recognition, and semantic labeling will benefit from and enable 'big data' repositories that intelligence agencies could use to expand intelligence gathering and analysis at tactical, operational, and strategic levels.[76] The US DoD, for example, is exploring the potential of AI for C2 systems, specifically to support multi-domain C2 missions.[77]

Recent technical studies generally agree that AI is an essential ingredient for fully autonomous systems – in both commercial and military contexts.[78] Various types of autonomous vehicles will rely on AI for route planning, perception, performing tactical maneuvers, recognizing obstacles, fusing sensor data, and communicating with other autonomous vehicles (i.e. during 'swarming' operations).[79] Great military powers have already invested heavily in developing dual-use autonomous technology, especially robotics and drones (see chapter 6).[80]

Even if AI cannot make better battlefield decisions than humans, militaries that use AI in human–machine teaming will doubtless gain significant advantages (remote-sensing, situational-awareness, cybersecurity, battlefield-maneuvers, and a compressed decision-making loop), compared to those who depend on human judgment, or semi-autonomous technology, alone[81] – in particular, in operating environments that demand endurance and rapid decision-making across multiple domains and combat zones.[82] According to the DoD's 2018 debut AI strategy report, "AI can help us better maintain our equipment, reduce operational costs, and improve readiness … [and] reduce the risk of civilian casualties and other collateral damage."[83]

Moreover, in complex and chaotic environments where humans are predisposed to use mental shortcuts to reduce the cognitive burden caused by information overload,[84] AI might greatly improve commanders' situational awareness, possibly giving them more time to make decisions (see chapter 8).[85] Theoretically, for example, AI deep neural networks could be used to enhance communication systems for classifying interference and optimizing spectrum usage. The salience of time in modern warfare is reflected in China's doctrinal *The Science of Military Strategy*, which emphasizes the role technology can play in reducing the gap in the "awareness-decision making operation" – akin to the observe, orient, decide, act decision-making loop.[86]

## Conclusion

This chapter defined and categorized the current state of AI (and critical AI-enabling technologies), and the potential limitations of this disruptive dual-use technology. It described the potential uses and implications of specific 'narrow' AI applications in the military arena, especially those that might impinge (directly or indirectly) on the nuclear domain. The chapter found that military AI is more akin to electricity, radios, radar, and military support systems than a stand-alone weapon as a constellation of technologies and capabilities.

However, AI suffers from several technical shortcomings that should prompt prudence and restraint in any decision to implement AI in the military – and especially a nuclear – context: a limited volume of quantity data-sets for ML algorithms to learn from; the inherent technical limitations of AI systems operating in a priori, complex, and adversarial environments; and the threat of adversarial AI exploitation and manipulation might prevent AI from reliably augmenting autonomous weapon systems (Part III will examine the strategic implications of these limitations). In sum, decision-makers must have a clear understanding of what AI is (and what it is not), its shortcomings (i.e. bias, unpredictability, and vast amounts of quality data-sets to learn and train algorithms) to prevent the premature adoption of unverified, accident-prone AI-augmented weapons.

The second section examined the significant gap between the reality of what AI technology is capable of doing in a military context, and the expectations and fears of public opinion, policy-makers, and global defense communities. It found that AI and autonomy could intersect with nuclear security in a multitude of ways, with both positive and negative implications for strategic stability. The use of AI at a strategic level of warfare (i.e. planning and directing future wars for the pursuit of political objectives), is a far less likely prospect in the near-term compared to supporting tactical and operational missions.[87]

Even in the absence of major scientific breakthroughs in AI, however, incremental steps along the path already established may still have potentially profound implications for strategic stability. Unfortunately, today, software engineers, military, and political leaders, and operators have only just begun to seriously consider the potential risks and trade-offs of employing these transformative enabling technologies safely and reliably in a military context, especially in the safety-critical nuclear domain.

The remainder of this book sketches a sobering portrait of these risks and trade-offs. How might states' pursuit of (offensive and defensive) advanced capabilities enhanced by AI technology create new uncertainties that exacerbate levels of mistrust and strategic competition between great military powers (chapter 3)? In what kinds of political regimes (i.e. democratic vs. non-democratic) might fears of militaries about the safety of AI-augmented systems likely trump the uncertainties of political leaders about the intentions of similarly equipped adversaries (chapter 8)? The next chapter conceptualizes AI with strategic stability in the second nuclear age and analyzes its impact on nuclear deterrence's central pillars.

## Notes

1 This chapter is intended to be an introductory primer to 'military AI' for non-technical audiences. It can only provide a snapshot in time because of the rapidly

developing nature of this field. The underlying AI-related technical concepts and analysis described in this primer will likely remain applicable in the near-term.
2 AI and machine learning (ML) are frequently used interchangeably. The two concepts are, in fact, technically distinct. ML is one particular method (or subset of AI technology) used to power AI systems (see Figure 1.2). However, because of the overwhelming success of ML algorithms compared to other methods, many AI systems today are based entirely on ML techniques.
3 A distinction is often made between automatic, automated, and autonomous systems, while others use these terms interchangeably. For this book, it is sufficient to acknowledge that the debate exists.
4 Andy Pearl, "Homage to John McCarthy, the Father of Artificial Intelligence (AI)," *Artificial Solutions*, June 2, 2017, www.artificial-solutions.com/blog/homage-to-john-mccarthy-the-father-of-artificial-intelligence (accessed February 20, 2021).
5 Nils J. Nilsson, *The Quest for Artificial Intelligence: A History of Ideas and Achievements* (Cambridge: Cambridge University Press, 2010).
6 Executive Office of the President, National Science and Technology Council, Committee on Technology, *Preparing for the Future of Artificial Intelligence*, October 12, 2016, https://obamawhitehouse.archives.gov/sites/default/files/whitehouse_files/microsites/ostp/NSTC/preparing_for_the_future_of_ai.pdf, p. 6 (accessed August 10, 2019).
7 Though many AI systems rely on large amounts of data, AI does not necessarily entail the volume, velocity, and variety usually associated with 'big data' analytics. Thus, while 'big data' is not new, and does not necessarily incorporate AI, AI techniques will likely supercharge the mining of data-rich sources. E. Gray et al., "Small Big Data: Using Multiple Data-Sets to Explore Unfolding Social and Economic Change," *Big Data & Society* 2, 1 (2015), pp. 1–6.
8 'Machine learning' is a concept that encompasses a wide variety of techniques designed to identify patterns in, and 'learn' and make predictions from, datasets. Successful 'learning' depends on having access to vast pools of reliable data about past behavior and successful outcomes. The 'neural network' approach to AI represents only a small segment of the improvements in AI techniques. AI also includes, for example, language processing, knowledge representation, and inferential reasoning, which is being actualized by the rapid advancements in software, hardware, data collection, and data storage. Jürgen Schmidhuber, "Deep learning in neural networks: an overview," *Neural Networks*, 61 (January 2015), pp. 85–117.
9 US technology companies reportedly invested an estimated US$20–US$30 billion in narrow AI algorithms in 2016, and this amount is expected to reach US$126 billion by 2025. Daniel S. Hoadley and Nathan J. Lucas, *Artificial Intelligence and National Security* (Washington, DC: Congressional Research Service, 2017), https://fas.org/sgp/crs/natsec/R45178.pdf, p. 2 (accessed August 10, 2019).
10 See Stuart Russell and Peter Norvig, *Artificial Intelligence: A Modern Approach*, 3rd ed. (Harlow: Pearson Education, 2014); and Nilsson, *The Quest for Artificial Intelligence*.

11 During the 'learning' process, machine learning algorithms generate statistical models to accomplish a specified task in situations it has not previously encountered. Most recent advances in AI applications have occurred with 'non-symbolic' ML approaches – computationally intensive, linear algebra, and statistical methods where input and knowledge representation are numerical (e.g. image pixels and audio frequencies. See Russell and Norvig, *Artificial Intelligence: A Modern Approach*.
12 AI 'narrow' breakthroughs have led to speculation about the arrival of AGI. Stuart Armstrong, Kaj Sotala, and Seán S. ÓhÉigeartaigh, "The Errors, Insights, and Lessons of Famous AI Predictions – and What they Mean for the Future," *Journal of Experimental & Theoretical Artificial Intelligence*, 26, 3 (2014), pp. 317–342.
13 Technology for Global Security (T4GS), "AI and Human Decision-Making: AI and the Battlefield," *T4GS Reports*, November 28, 2018, www.tech4gs.org/ai-and-human-decisionmaking.html (accessed August 10, 2019).
14 Stuart Russell, *Human Compatible: Artificial Intelligence and the Problem of Control* (New York: Viking Press, 2019), p. 136.
15 For an excellent overview of AI and its various subfields, see Margaret A. Boden, *AI: Its Nature and Future* (Oxford: Oxford University Press, 2016); and David Vernon, *Artificial Cognitive Systems: A Primer* (Cambridge, MA: MIT Press, 2014).
16 'Narrow' AI has inherent limitations and lacks a capacity for will and intent. Used in a military context it is, therefore, a tool to enhance or enable weapon systems, not an independent actor. See Lawrence Lewis, *AI and Autonomy in War* (Washington, DC: Center for Naval Analysis, August 2018), www.cna.org/CNA_files/PDF/Understanding-Risks.pdf, p. 17 (accessed February 20, 2021).
17 Historical definitions of AI can be grouped as follows: systems that think like humans; systems that act like humans; and systems that act and reason. See Andrew W. Moore, "AI and National Security in 2017," Presentation at AI and Global Security Summit, Washington, DC, November 1, 2017; and Andrew Ilachinski, *AI, Robots, and Swarms: Issues, Questions, and Recommended Studies* (Washington, DC: Center for Naval Analysis, January 2017), p. 6.
18 The term revolution in military affairs was popularized in the 1990s and early 2000s, especially within the US defense community. An RMS is usually caused by technology but is not *revolutionary* unless it changes how states fight wars, instigating, for example, new operational concepts, tactics, and altering military balances of powers. See MacGregor Knox and Williamson Murray, *The Dynamics of Military Revolution* (Cambridge: Cambridge University Press, 2001), pp. 1300–2050; and Colin S. Gray, *Strategy for Chaos: Revolutions in Military Affairs and the Evidence of History* (London: Frank Cass, 2002). For Chinese views on the debate see Fu Wanjuan, Yang Wenzhe, and Xu Chunlei, "Intelligent Warfare, Where Does it Not Change?" *PLA Daily*, January 14, 2020, www.81.cn/jfjbmap/content/2020-01/14/content_252163.htm; and Lin Juanjuan, Zhang Yuantao, and Wang Wei "Military Intelligence is Profoundly Affecting Future Operations," *Ministry of National Defense of the People's*

*Republic of China*, September 10, 2019, www.mod.gov.cn/jmsd/201909/10/content_4850148.htm (accessed February 20, 2021).
19 Hoadley and Lucas, *Artificial Intelligence and National Security*, pp. 1–2.
20 US Department of Defense, "Summary of the 2018 Department of Defense Artificial Intelligence Strategy," https://media.defense.gov/2019/Feb/12/2002088963/-1/-1/1/SUMMARY-OF-DOD-AI-STRATEGY.PDF (accessed August 10, 2019).
21 Ibid., p. 3; Moore, "AI and National Security in 2017."
22 'Autonomy' in the context of military applications can be defined as the condition or quality of being self-governing to achieve an assigned task, based on a system's situational awareness (integrated sensing, perceiving, and analyzing), planning, and decision-making. An autonomous weapon system (or lethal autonomous weapon system) is a weapon system that, once activated, can select and engages targets without further intervention by a human operator. See "Autonomy in Weapon Systems," US Department of Defense, *Directive Number 3000.09*, May 8, 2017, www.esd.whs.mil/Portals/54/Documents/DD/issuances/dodd/300009p.pdf (accessed September 10, 2019).
23 Michael C. Horowitz, "Artificial Intelligence, International Competition, and the Balance of Power," *Texas National Security Review*, 1, 3 (2018), pp. 37–57.
24 See Nadine Sarter, David Woods, and Charles Billings, "Automation surprises," in G. Salvendy (ed.), *Handbook of Human Factors and Ergonomics*, 2nd ed. (New York: John Wiley & Sons Inc. 1997).
25 AI systems are not merely collections of hardware and algorithms but, rather, supervisory control systems. As such, these systems need to be observable, explainable, controllable, and predictable, and to direct the attention of their human supervisors when required by a particular situation or operation. David Woods and Erik Hollnagel, *Joint Cognitive Systems: Patterns in Cognitive Systems Engineering* (London: Routledge, 2006), pp. 136–137.
26 Osob A. Osonde and William Welser IV, *An Intelligence in Our Image: The Risks of Bias and Errors in Artificial Intelligence* (Santa Monica, CA: RAND Corporation, 2017), www.rand.org/pubs/research_reports/RR1744.html (accessed February 20, 2021).
27 Stephen Van Evera, *Causes of War: Power and the Roots of Conflict* (Ithaca, NY: Cornell University Press, 1999).
28 Jason Matheny, Director of Intelligence Advanced Research Program Agency, as cited in Hoadley and Lucas, *Artificial Intelligence and National Security*, p. 9.
29 Matthew Hutson, "AI Researchers Allege That Machine Learning is Alchemy," *Science*, May 3, 2018, www.sciencemag.org/news/2018/05/ai-researchers-allege-machine-learning-alchemy.
30 The concept of 'learning' in the context of ML refers to finding statistical relationships in past data, as opposed to anthropomorphic interpretations that confuse human learning with machine 'learning.' See David Watson, "The Rhetoric and Reality of Anthropomorphism in Artificial Intelligence," *Minds and Machines*, 29 (2019), pp. 417–440.
31 A computational method known as 'artificial neural networks' draws on knowl-

edge of the human brain, statistics, and applied mathematics. Nilsson, *The Quest for Artificial Intelligence*, chs 28–29.
32  Deep learning is an approach to ML whereby the system 'learns' how to undertake a task in either a supervised, semi-supervised, or unsupervised manner. Ian Goodfellow, Yoshua Bengio, and Aaron Courville, *Deep Learning* (Cambridge, MA: MIT Press, 2016).
33  Several recent AI innovations continue to use the previous generation of hard-coded techniques such as Bayesian statistics, probabilistic relational models, and other evolutionary algorithms. Nilsson, *The Quest for Artificial Intelligence*, chs 28–29.
34  Ibid., p. 347.
35  In contrast, hand-coded programming often requires a significant amount of research on how the world works – or contextual analysis. Russell and Norvig, *Artificial Intelligence: A Modern Approach*, chapter 18.
36  Lewis, *AI and Autonomy in War*.
37  Juanjuan, Yuantao, and Wei, "Military Intelligence is Profoundly Affecting Future Operations."
38  Lt. General John Shanahan, Director of the DoD JAIC, quoted in, Sydney J. Freedberg Jr., "EXCLUSIVE Pentagon's AI Problem is 'Dirty' Data: Lt. Gen. Shanahan," *Breakingdefense*, November 13, 2019, https://breakingdefense.com/2019/11/exclusive-pentagons-ai-problem-is-dirty-data-lt-gen-shanahan/ (accessed December 10, 2019).
39  Ian Goodfellow, Patrick McDaniel, and Nicolas Papernot, "Making Machine Learning Robust Against Adversarial Inputs," *Communications of the ACM*, 61, 7 (2018), pp. 56–66; and Ian Goodfellow, Jonathon Shlens, and Christian Szegedy, "Explaining and Harnessing Adversarial Examples," December 20, 2014, *arXiv*, https://arxiv.org/pdf/1412.6572.pdf (accessed February 20, 2021).
40  As variance decreases, bias usually increases. See Pedro Domingos, "A Few Useful Things to Know About Machine Learning," *Communications of the ACM*, 55, 10 (2012), pp. 85–86.
41  Less sophisticated AI systems will exhibit greater levels of bias, but with lower accuracy.
42  Recent research in adversarial AI has discovered blind spots, and a minimal understanding of the context in a fast-moving and complex environment – or 'brittleness.' See Goodfellow, Shlens, and Szegedy, "Explaining and Harnessing Adversarial Examples."
43  Joseph Johnson, "MAD in an AI Future?" Center for Global Security Research (Livermore, CA: Lawrence Livermore National Laboratory), pp. 4–6.
44  As 'unsupervised' ML techniques mature, however, the reliance on data and labeling (i.e. images, videos, and text) to support AI system's training environments is expected to decrease. As opposed to current 'supervised' (or reinforcement) learning techniques that depend on labeling images to detect patterns, unsupervised machine-learning is designed to create autonomous AI by rewarding agents (i.e. ML algorithms) for learning about the data they observe without a particular task in mind. See Alexander Graves and Kelly Clancy, "Unsupervised

Learning: The Curious Pupil," *Deepmind,* June 25, 2019, https://deepmind.com/blog/article/unsupervised-learning (accessed December 10, 2019).

45 When solving problems, humans learn high-level concepts with relatively little data, which apply to other problems. AI does not possess this broader knowledge or common sense. Goodfellow, Shlens, and Szegedy, "Explaining and Harnessing Adversarial Examples;" and Brenden M. Lake, Ruslan Salakhutdinov, and Joshua B. Tenenbaum, "Human-Level Concept Learning Through Probabilistic Program Induction," *Science,* 350, 6266 (2015), pp. 1332–1338.

46 AI systems are limited as to what they can infer from particular data-sets because of the relatively few higher-level mathematical concepts on which computational-learning theory is derived. David Deutsch, "Creative Blocks," *Aeon,* October 3, 2012, https://aeon.co/essays/how-close-are-we-to-creating-artificial-intelligence (accessed December 10, 2019).

47 Domingos, "A Few Useful Things to Know About Machine Learning," pp. 78–88.

48 Technical advances in decision trees, support vector machines, and deep learning have enabled a greater level of flexible discriminators into AI, but they still exhibit similar learning issues as the resolution and dimensions of images increases. For a study on the nature of human cognition see Daniel Kahneman, *Thinking, Fast And Slow* (New York: Penguin 2011).

49 Deutsch, "Creative Blocks."

50 Leslie Kaelbling, Michael L. Littman, and Anthony R. Cassandra, "Planning and Acting in Partially Observable Stochastic Domains," *Artificial Intelligence,* 10, 3 (June 2017), pp. 99–134.

51 The first (from 1760 to 1840), brought the steam engine, railroads, and machine manufacturing. The second (from 1870 to 1914), gave us electricity and mass production. The third, often called the digital or computer revolution (in the last decades of the twentieth century), produced semiconductors, computers, and the internet. David Barno and Nora Bensahel, "War in the Fourth Industrial Revolution," *War on the Rocks,* July 3, 2018, https://warontherocks.com/2018/06/war-in-the-fourth-industrial-revolution/ (accessed December 10, 2019).

52 Notable exceptions include: Vincent Boulanin (ed.), *The Impact of Artificial Intelligence on Strategic Stability and Nuclear Risk Vol. I Euro-Atlantic Perspectives* (Stockholm: SIPRI Publications, May 2019); Edward Geist and Andrew Lohn, *How Might Artificial Intelligence Affect the Risk of Nuclear War?* (Santa Monica, CA: RAND Corporation, 2018); Michael Horowitz, Paul Scharre, and Alex Velez-Green, *A Stable Nuclear Future? The Impact of Automation, Autonomy, and Artificial Intelligence* (Philadelphia, PA: University of Pennsylvania Press, 2017); and James S. Johnson, "Artificial Intelligence: A Threat to Strategic Stability," *Strategic Studies Quarterly,* 14, 1 (2020), pp. 16–39.

53 Some neuroscientists have suggested that the exaggerated threat narrative surrounding AI can be attributed to the way humans, from an evolutionary perspective, tend to conflate intelligence with the drive to achieve dominance.

Anthony Zador and Yann LeCun, "Don't Fear the Terminator," *Scientific American*, September 26, 2019, https://blogs.scientificamerican.com/observations/dont-fear-the-terminator/ (accessed December 10, 2019). For example, George Zarkadakis, *In Our Image: Savior or Destroyer? The History and Future of Artificial Intelligence* (New York: Pegasus Books, 2015); Christianna Ready, "Kurzweil Claim that Singularity will Happen by 2045," *Futurism*, October 5, 2017, https://futurism.com/kurzweil-claims-that-the-singularity-will-happen-by-2045 (accessed March 10, 2020).

54 Boulanin (ed.), *The Impact of Artificial Intelligence on Strategic Stability and Nuclear Risk Vol. I Euro-Atlantic Perspectives*, pp. 28–30.
55 Connie Lee, "SOCOM Plans New Artificial Intelligence Strategy," *National Defense*, August 9, 2019, www.nationaldefensemagazine.org/articles/2019/8/9/socom-plans-new-artificial-intelligence-strategy (accessed February 20, 2021).
56 Aerial and under-water drones in swarms could eventually replace ICBMs and nuclear-powered ballistic missile submarines (SSBNs) for the delivery of nuclear weapons.
57 For example, in 2017 China's Institute of Command and Control held the nation's first-ever "Artificial Intelligence and War-Gaming National Finals." Chinese leaders believe these games will allow their personnel to gain a greater appreciation of the trends in warfare, especially given the lack of recent combat experience.
58 For example, AI is enabling scientists to model nuclear effects to confirm the reliability of the nuclear stockpile without nuclear testing. See *Strategic Latency and Warning: Private Sector Perspectives on Current Intelligence Challenges in Science and Technology*, Report of the Expert Advisory Panel Workshop, Lawrence Livermore National Laboratory, January 8, 2016, https://e-reports-ext.llnl.gov/pdf/804661.pdf (accessed December 10, 2019).
59 For example, the US National Geospatial-Intelligence Agency has reportedly used AI to support military and intelligence analysis. Ben Conklin, "How Artificial Intelligence is Transforming GEOINT," *GCN*, April 18, 2018, https://gcn.com/articles/2018/04/18/ai-transform-geoint.aspx (accessed December 10, 2019).
60 For example, the US Air Force is currently developing AI tools to accomplish simple predictive aircraft maintenance, and a Texas-based AI company, SparkCognition, recently installed AI predictive technology in a Boeing commercial aircraft. Marcus Weisgerber, "Defense Firms to Air Force: Want Your Planes' Data? Pay Up," *Defense One*, September 19, 2017, www.defenseone.com/technology/2017/09/military-planes-predictive-maintenancetechnology/141133/ (accessed December 10, 2019).
61 Scott Rosenberg, "Firewalls Don't Stop Hackers, AI might," *Wired*, August 27, 2017, www.wired.com/story/firewalls-dont-stop-hackers-ai-might/ (accessed December 10, 2019).
62 "Mayhem Declared Preliminary Winner of Historic Cyber Grand Challenge," August 4, 2016, www.darpa.mil/news-events/2016-08-04 and http://archive.darpa.mil/cybergrandchallenge/ (accessed December 10, 2019).
63 Several states, notably China and Russia, are researching the use of machine

learning to develop control systems for hypersonic vehicles; these capabilities cannot be operated manually because of their high velocity.
64 Ben Connable, *Embracing the Fog of War: Assessment and Metrics in Counterinsurgency* (Santa Monica, CA: RAND Corporation, 2012).
65 For example, AI used in conjunction with autonomous mobile sensor platforms might compound the threat posed to the survivability of mobile ICBM launchers. See Paul Bracken, "The Cyber Threat to Nuclear Stability," *Orbis*, 60, 2 (2016), pp. 188–203, p. 194.
66 A good case in point of a sensing error occurred in 2015 when Google's automated image recognition system in a photo application misidentified African Americans, creating an album titled 'Gorillas.' Amanda Schupak, "Google Apologizes for Mis-Tagging Photos of African Americans," *CBS News*, July 1, 2015, www.cbsnews.com/news/google-photos-labeled-pics-of-african-americans-as-gorillas/ (accessed December 10, 2019).
67 Keith Dear, "Artificial Intelligence and Decision-Making," *The RUSI Journal*, 164, 5–6 (2019), pp. 18–25.
68 Despite the rapid advances in AI research, experts remain in the dark about many aspects of the precise workings of neural networks work, how to improve their accuracy (besides just feeding them more data), or how to fix the biases that exist. Osoba and Welser, *An Intelligence in Our Image*.
69 Rodrick Wallace, *Carl von Clausewitz, the Fog-of-War, and the AI Revolution* (New York: Springer, 2018), pp. 1–45.
70 Colin S. Gray, "Why Strategy is Difficult," *Joint Forces Quarterly* (Summer 1999), pp. 7–12, p. 9.
71 Since the 1980s, uncertainty has been a recurring polemic in the context of 'modern AI' – and a ubiquitous issue in real-world strategic decision-making. Russell, *Human Compatible*, p. 176.
72 See Robert Jervis, *Perception and Misperception in International Politics* (Princeton, NJ: Princeton University Press, 1976); and Tarhi-Milo Keren, *Knowing the Adversary: Leaders, Intelligence Organizations, and Assessments of Intentions in International Relations* (Princeton, NJ: Princeton University Press, 2015).
73 Target intelligence that relies on pattern recognition software (especially against mobile missile targets) has been the biggest obstacle to effective counterforce. US Cold War-era efforts to target Soviet mobile ICBM launchers combined bombardment strategies with intelligence that searched for patterns in the way the Soviet Union moved its missiles. Geist and Lohn, *How Might Artificial Intelligence Affect the Risk of Nuclear War?* p. 16. For recent Chinese views on AI's implications for intelligence see Juanjuan, Yuantao, and Wei, "Military Intelligence is Profoundly Affecting Future Operations."
74 Also, the US Central Intelligence Agency (CIA) has 137 projects in development that leverage AI in some capacity to accomplish tasks such as image recognition or labeling (similar to Project Maven's algorithm and data analysis functions) to predict future events like terrorist attacks or civil unrest based on wide-ranging analysis of open-source information. Gregory Allen, "Project Maven Brings AI to the fight against ISIS," *The Bulletin of the Atomic Scientists* (December 21,

2017), https://thebulletin.org/project-maven-brings-ai-fight-against-isis11374. (accessed December 10, 2019).

75  Defense experts estimate that while AI will solve some of these challenges in the near-term (i.e. within five years), numerous technical problems associated with tracking and targeting mobile missiles are unlikely to be overcome within the next two decades. Even with perfect knowledge of the target location, for example, mobile targets can move between the time a weapon is launched and the time it arrives. Ibid., pp. 16–17.

76  For example, China recently deployed a software package called *Xiu Liang* ('Sharp Eyes') to identify criminals in housing complexes and public areas through a combination of face recognition and CCTV footage. Simon Denyer, "In China, Facial Recognition is Sharp End of a Drive for Total Surveillance," *Chicago Tribune*, January 8, 2018, www.washingtonpost.com/news/world/wp/2018/01/07/feature/in-china-facial-recognition-is-sharp-end-of-a-drive-for-total-surveillance/?utm_term=.fcd7bdaa24a9 (accessed December 10, 2019).

77  Mark Pomerlau, "How Industry's Helping the US Air Force with Multi-Domain Command and Control," *Defense News*, September 25, 2017, www.defensenews.com/c2–comms/2017/09/25/industry-pitches-in-to-help-air-force-with-multi-domain-command-and-control/ (accessed December 10, 2019).

78  AI techniques for these functions are similar to those for commercial self-driving vehicles. Bill Canis, *Issues in Autonomous Vehicle Deployment* (Washington, DC: Congressional Research Service, 2017), pp. 2–3.

79  AI-enabled drone swarming (or cooperative behavior) is a subset of autonomous vehicle development. Ilachinski, *AI, Robots, and Swarms*, p. 108.

80  I. Sutyagin, "Russia's Underwater 'Doomsday Drone:' Science Fiction, But Real Danger," *Bulletin of the Atomic Scientists*, 72, 4 (June 2016), pp. 243–246; Mary-Ann Russon, "Google Robot Army and Military Drone Swarms: UAVs May Replace People in the Theatre of War," *International Business Times*, April 16, 2015, www.ibtimes.co.uk/google-robot-army-military-drone-swarms-uavs-may-replace-people-theatre-war-1496615 (accessed February 20, 2021); and Elsa Kania, *Battlefield Singularity: Artificial Intelligence, Military Revolution, and China's Future Military Power* (Washington, DC: Center for a New American Security, November 2017), p. 23.

81  In contrast to human decision-makers, cognitive stressors, time pressures, and other physical effects of combat (such as lack of glucose and fatigue), do not adversely affect AI systems. Kareem Ayoub and Kenneth Payne, "Strategy in the Age of Artificial Intelligence," *Journal of Strategic Studies* 39, 5–6 (2016), pp. 793–819.

82  The CIA is actively pursuing several publicly documented AI research projects to reduce the 'human-factors burden,' increase actionable military intelligence, enhance military decision-making, and ultimately, to predict future attacks and national security threats. Patrick Tucker, "What the CIA's Tech Director Wants from AI," *Defense One*, September 6, 2017, www.defenseone.com/technology/2017/09/cia-technology-director-artificial-intelligence/140801/. (accessed December 10, 2019).

83 Memorandum from the Deputy Secretary of Defense, "Establishment of the Joint Artificial Intelligence Center," June 27, 2018; https://admin.govexec.com/media/establishment_of_the_joint_artificial_intelligence_center_osd008412-18_r....pdf (accessed December 10, 2019).
84 Daniel Kahneman and Shane Frederick, "Representativeness revisited: attribute substitution in intuitive judgment," inTimothy Gilovich, Daniel Griffin, and Daniel Kahnema (eds), *Heuristics and Biases: The Psychology of Intuitive Judgment* (Cambridge: Cambridge University Press, 2002), pp. 49–81.
85 Yang Feilong and Li Shijiang "Cognitive Warfare: Dominating the Era of Intelligence," *PLA Daily*, March 19, 2020, www.81.cn/theory/2020-03/19/content_9772502.htm (accessed December 10, 2019).
86 Research Department of Military Strategy, *The Science of Military Strategy*, 3rd ed. (Beijing: Military Science Press, 2013), p. 189. Also see Yang and Li "Cognitive Warfare."
87 For a recent study of the ambiguities and inconsistencies in the way AI researchers conceptualize the non-binary relationship between near- and long-term AI progress, see Carina Prunkl and Jess Whittlestone, "Beyond Near- and Long-Term: Towards a Clearer Account of Research Priorities in AI Ethics and Society," in Proceedings of the 2020 AAAI/ACM Conference on AI, Ethics, and Society (AIES 2020), February 2020, https://doi.org/10.1145/3375627.3375803 (accessed March 10, 2020).

# 2

# AI in the second nuclear age

How can we best conceptualize AI and military technological change in the context of nuclear weapons? Despite being theoretically and politically contested to this day, the notion of 'strategic stability' has proven a useful intellectual tool for analyzing the potential for new, powerful, and technically advanced weapons to undermine stability between nuclear-armed adversaries.[1] The concept entered into the nuclear lexicon during the early 1950s and is inextricably connected to the strategic thinking and debates that surrounded the 'nuclear revolution,'[2] including: how a nuclear war might be fought; the requirements and veracity of credible deterrence; the potential risks posed by pre-emptive and accidental strikes;[3] and how to ensure the survivability of retaliatory forces.[4] In short, strategic stability provides an over-arching theoretical framework for understanding the nature of security in the nuclear age (i.e. evaluating nuclear force structures, deployment decisions, and a rationale for arms control).[5]

During the twentieth century, the world eventually found a way to manage a paradigm shift in military technology – the atomic bomb. Nuclear warfare was avoided through a combination of deterrence, arms control, and safety and verification measures. In the second nuclear age, states continue to face many of the same challenges and trade-offs in the use of advanced technology in the nuclear domain that the Cold War warriors wrestled with, inter alia:[6] (1) effective international regulation of atomic weapons; (2) safe and reliable military-civilian control of nuclear weapons and their attendant systems; (3) the prospect of nuclear blackmail of non-nuclear rivals; (4) the perceived risk of disarming pre-emptive strikes by adversaries in asymmetric situations; (5) a shifting strategic military balance between nuclear-armed adversaries; (6) the use of nuclear weapons in warfighting doctrine; (7) the use of nuclear weapons in the context of conventional conflict; and (8) striking an appropriate balance between signaling resolve and deterrence, and escalating a situation.[7]

This chapter proceeds in two sections. The first offers a brief literature review on the concept of 'strategic stability,' and the relevance of this Cold

War term for conceptualizing emerging technological challenges to strategic stability in the second nuclear age. It argues that the goal of preserving stability to prevent a nuclear war that emerged during the Cold War has implications for today's discourse on AI and strategic stability. The second section contextualizes AI within the broad spectrum of military technologies associated with the 'computer revolution.' It argues that military AI and the advanced capabilities it enables are best viewed as a natural manifestation of an established trend in emerging technology. Even if AI does not become the next revolution in military affairs, it could have significant implications for the central pillars of nuclear deterrence.

## Strategic stability: a platonic ideal?

Combining cognition, stress, strategic culture, wargaming, and game theory, strategic stability provided fertile ground for scholars who, by the early 1960s, began to incorporate the term into their research.[8] In the broadest use of the term, strategic stability exists in the absence of a significant incentive for an adversary to engage in provocative behavior.[9] In other words, there was a lack of armed conflict and perceived incentive to use nuclear weapons first between nuclear-armed states.[10] At its core, the concept is a dynamic and fluid one that focuses on finding a *modus vivendi* for the complex interactions and incentives of two (or more) actors.[11] The term is often associated with the relative power distribution among great and rising powers, particularly those in possession of nuclear weapons or the potential to acquire them. Strategic stability is ultimately a product of a complex interplay of political, economic, and military dynamics in which technology performs several functions – an equalizer, counterweight, and principal agent of change.[12]

Technology has long been used to augment, automate, and enhance human behavior and decision-making in a military context; the 'human factor' has, until now, trumped technology in its impact on strategic stability.[13] That is, the underlying forces behind fundamental shifts in strategic stability are generally less concerned with quantitative or qualitative assessments of military capabilities (and other measures of relative power), and are, instead, more focused on how nuanced institutional, cognitive, and strategic variables impact strategic decision-making and may cause misperceptions of others' intentions.[14] 'Stability' is concerned with the relationship between these factors, particularly what increases their capabilities and for what purpose.[15] The role of technological change and strategic stability can be conceptualized, therefore, as part of a complex interaction of disruptive forces (or agents of change), which during periods of heightened geopolitical rivalry, great-power transitions, and strategic surprise, may erode strategic stability and make conflict more likely.[16]

The military-technical historical record attests that uncertainty caused by the perceived utility (i.e. its speed, scope, the rate of proliferation, and diffusion) of a new application significantly outweighs its *actual* technical feasibility as a potential source of strategic competitiveness.[17] For example, the Sputnik satellite launch becomes analogous with the outsized impact a seemingly simple technological demonstration can have on strategic stability (see chapter 3). A core argument of this book is that AI's impact on states' nuclear strategy will ultimately depend as much (or perhaps more so) on adversaries' perceptions of its capabilities than on what a particular AI-enabled application is technically capable of doing.[18]

Three broad forms distinguish strategic stability: first-strike stability; crisis stability; and arms-race stability. *First-strike stability* exists in situations when no one state can launch a surprise (or pre-emptive) attack against an opponent without the fear of devastating reprisals from survivable second-strike forces. That is, the lack of *both* incentives or pressures to use nuclear weapons first in a crisis.[19] Thus, fear that the advantage of first-strike capabilities could be eroded or neutralized by an adversary would be destabilizing and could increase incentives to launch a pre-emptive strike.[20] For example, throughout the Cold War, as today, the vulnerability of command and control structures to counterforce attack remains high and is compounded by the increasing complexity of these systems, and the propensity of states to use them to support both nuclear and non-nuclear forces.

The rapid proliferation, diffusion, and ubiquity of advanced technologies like offensive cyber, hypersonic weapons, and AI and autonomous weapon systems will make it increasingly difficult for states to mitigate this vulnerability without simultaneously improving their ability to strike first, thereby undermining the survivability of others' strategic forces (see chapters 4 and 8).[21] A crucial distinction, therefore, is between the risk of unintentional (i.e. accidental or inadvertent) escalation and intentional escalation; the fear of deliberate escalation is generally more destabilizing.[22] Ultimately, whether the impact of unintended escalation risk is stabilizing or destabilizing depends on the relative strengths of the destabilizing force and the fear it instills.[23] Measures to promote first-strike stability are concerned with reducing use-them-or-lose-them incentives, and the risk of inadvertent escalation.

*Crisis stability* aims to prevent (or deescalate) escalation during crises – such as those that occurred in Berlin and Cuba in the early 1960s.[24] Crisis stability, therefore, depends on reciprocal fear between states; when crises do arise, the system does not worsen the situation. Conversely, crisis instability refers to what Thomas Schelling called the "reciprocal fear of surprise attack."[25] That is, the belief that conflict is inevitable, and thus striking first and pre-emptively would create a strategic advantage.[26] Finally, *arms-race stability* (during peacetime) can emerge when there are no exploitable inequalities (or asymmetries) separating adversaries' military forces – qualitatively or quantitatively.[27] Thus, a relatively weak adversary will have a strong incentive to modernize its strategic nuclear forces to reduce their vulnerability to pre-emptive attacks, and improve its ability to penetrate the other side's strategic defenses.[28]

Crisis instability and arms-race instability can arise when states use strategic capabilities and instill fear into a situation.[29] This fear closely correlates with the incentives these decisions create. Thus, if both states possess first-strike capabilities, the incentive for either side to gain by choosing to strike first depends on the fear that hesitation might allow a rival to gain the upper hand.[30] In other words, *both* the incentive and fear created by strategic weapons can exacerbate escalation risk.[31] In this way, crisis (in)stability is fundamentally a psychological problem.[32] Similarly, in an arms race, the incentive to gain the upper hand vis-à-vis a rival is correlated with the fear that not doing so will put an adversary in an advantageous position. Because of this connection, strategic stability is often characterized as crisis stability plus arms-race stability. Given the natural tension and trade-offs between these objectives, calibrating an effective nuclear strategy in bargaining situations – especially between deterrence and assurance – has eluded policy-makers.[33]

Scholars Dale Walton and Colin Gray have argued that "true strategic stability is a platonic ideal, useful as a yardstick for judging real-world conditions, but inherently unattainable as a policy goal" – especially in a multipolar nuclear world order.[34] Though a stable strategic environment is not necessarily a catalyst for positive geopolitical change, a conflict between great powers is more likely when the current strategic environment is unstable and uncertain.[35] Political thought scholarship has demonstrated that security competition – motivated by the desire to control warfare – tends to be ratcheted up because of the complexity and uncertainty of military technology and operations overtime; akin to the Clausewitzian conditions of 'fog and friction.'[36]

Furthermore, we are unable to assume *ex-ante* that a specific policy (e.g. arms control or confidence-building measures) which might improve (or worsen) strategic stability will necessarily have equally positive (or negative) outcomes for the other variables involved in nuclear policy (e.g. credible deterrence, bureaucratic feasibility, alliance politics, and domestic political motives).[37] A strategy of offshore balancing could be used, for example, to restore the balance of power (i.e. stabilizing) within a regional alliance structure vis-à-vis a rising revisionist power. Conversely, bandwagoning pressures within the system may compound disruptive changes in the balance of power caused by this behavior and destabilization.

According to scholar Robert Jervis, the deterrence implications of invulnerable nuclear arsenals are "many and far-reaching."[38] That is, the existence of survivable nuclear arsenals (or second-strike capabilities) should ensure that war does not occur, crises will be rare, and the status quo will be relatively easy to maintain. Scholars have reconciled the persistently high levels of strategic competition and seemingly anomalous states' behavior during the nuclear era (i.e. arms racing, territorial struggles, and alliance building), with misguided (or mistaken) decisions by leaders, bureaucratic pathologies, and domestic political factors.[39] More recently, scholars have argued that the idea of a 'nuclear revolution theory' is itself a flawed intellectual construct.[40]

Since the introduction of nuclear weapons, there have been several technological threats to stability.[41] Keir Lieber and Daryl Press argue that global nuclear forces have become increasingly vulnerable due to rapid improvements in conventional counterforce technologies, explaining why strategic tensions have persisted.[42] Taken together, advances in technologies such as imagery and sensor technology, data processing, communications, unmanned vehicles, cyberspace, and AI have increased the uncertainty about states' future capabilities, both real and perceived. Strategic instability, therefore, involves a quantitative (i.e. capabilities) as well as a qualitative (i.e. intentions) assessment. Thus, nuclear powers must maintain a diverse field of cutting-edge nuclear and conventional weapon systems (both kinetic and non-kinetic) to hedge against the possibility of future technological breakthroughs (chapters 4 and 5 consider the nature and strategic impact of this hedge).[43]

Research on bargaining and war also suggests that uncertainty about states' capabilities makes it more challenging to reach diplomatic solutions once a crisis or conflict begins.[44] Further, the increasing co-mingling (or blurring) of the nuclear and non-nuclear domains since the end of the Cold War (e.g. nuclear launch and delivery structures, and their attendant command, control, communications, and intelligence networks) will continue to have destabilizing implications for nuclear escalation.[45] Thomas Schelling and Morton Halperin's Cold War-era arguments about emerging technology and new challenges to strategic stability and arms control resonate today:

> The present race seems unstable *because of the uncertainty in technology and the danger of a decisive breakthrough*. Uncertainty means that each side must be prepared to spend a great deal of money; it also means a constant fear on either side that the other has developed a dominant position, or will do so, or

will fear the first to do so, with the *resulting danger of premeditated or preemptive attack* (emphasis added).[46]

The goal of preserving stability to prevent catastrophic nuclear war and the instability that emerged during the Cold War has important implications for the (albeit lightly researched) discourse on AI and strategic stability. Four interrelated lessons for stability and advanced technology emerged from the Cold War experience. First, the destabilizing impact of weapons proliferation (both vertical and horizontal) becomes more acute in situations where technologies have the potential to create asymmetric military advantages between rival states such as: improved intercontinental ballistic missile accuracy; ballistic missiles armed with multiple independent targetable re-entry vehicles; ballistic anti-ship missiles; and missile defense technology. Technologies like these can incentivize states to diversify, secure, and harden their strategic forces.[47] Ballistic missiles, for example, caused a trade-off between the perceived utility of independently targeted re-entry vehicles to maximize the success (or perceived success) of a disabling first strike, and the detrimental impact of this asymmetry for strategic stability.[48]

Second, where the risk exists – or is perceived to exist – that a conventional conflict might cross the nuclear threshold, the implications for crisis instability and military escalation can be severe.[49] The US Office of Technology Assessment described this risk's nature in the following way:

> The degree to which strategic force characteristics might, in a crisis situation, reduce incentives to initiate the use of nuclear weapons. Weapon systems are *considered destabilizing* if, in a crisis, they would *add significant incentives to initiate a nuclear attack*, and particularly to attack quickly before there is much time to collect reliable information and carefully weigh all available options and their consequences (emphasis added).[50]

Third, any meaningful definition of strategic stability must include advanced strategic non-nuclear weapons – notably conventional counterforce capabilities – that adversaries could use to destroy an enemy's nuclear capabilities (e.g. ballistic missile defense, cyber, AI, anti-satellite weapons, anti-submarine weapons, and precision strike munitions).[51] That is, technologically advanced offensive and defensive non-nuclear weapons can reduce states' vulnerability to a nuclear attack.[52]

Finally, and related, nuclear modernization programs that are primarily motivated by the technological advancement (or technological determinism), in particular associated with increased levels of automation, often trigger arms-race dynamics that worsen crisis stability.[53] The intense bipolar military-technical strategic rivalry (both offensive and defensive) during the Cold War can give us a better sense of how the recent developments in military AI could affect future strategic stability.[54]

## AI and strategic stability: revisiting the analytical framework

The recent history of emerging technology and revolution in military affairs counsels caution in extrapolating from current (or past) technological trends.[55] Technology rarely evolves in the way futurists predict, and many defense innovations have had countervailing or conditional effects that failed to live up to exaggerated predictions, or were proved wrong.[56] With aircraft came radar and anti-aircraft guns; with submarines came anti-submarine warfare tactics such as depth charges; and with the development of poisonous chemical gas came gas masks.

Many predicted, for instance, that chemical weapons would instantly and dramatically change the nature of warfare and deterrence after the British used poison gas during World War I. However, chemical weapons proved far less practical, impactful, disruptive, as well as being relatively easier to defend against, than conventional explosives.[57] Similarly, inter-war strategists believed that the tank would revolutionize warfare. In the event, the German anti-tank weapon and modern warfare tactics were able to countervail British tanks during *Operation Goodwood* effectively. More recently, US 'network-centric warfare' did not prove to be the strategic game changing innovation predicted by many military thinkers.[58]

As a corollary, AI-powered drone swarms will likely be countervailed by parallel advances in defensive uses of image and remote sensor technology to detect and track swarms, and digital jammers to interfere with their networks' high-energy lasers to destroy them. Further, offensive cyber could also be used in 'left of launch' operations (see chapter 7) to prevent these swarms from mounting an attack in the first place.[59] Technological-deterministic schools of thought that conceptualize technology as the *sole* driver of strategic stability and the conduct of warfare oversimplify this complicated human endeavor.[60] As discussed in the opening chapter, even if AI ultimately has a transformative strategic effect, it will likely be far more non-linear, iterative, and prosaic.[61] As Bernard Brodie cautioned during the Cold War-era: "What may look like extraordinarily important changes in the tools of war or related technologies may appear to lack a significant impact on strategy and politics ... *even though those improvements may look quite significant* to a scientist or an engineer" (emphasis added).[62]

From this perspective, the increasing speed of warfare, shortening of the decision-making timeframe, and the co-mingling of military capabilities that have occurred within the broader context of emerging technology would likely have occurred whether or not AI was involved.[63] Put another way, AI and the advanced weapon systems it might augment can be viewed as a natural manifestation – rather than the cause or origin – of an established trend in emerging technology, potentially leading states to adopt destabilizing launch postures.[64] These trends also demonstrate the difficulty of applying traditional definitions of strategic stability to advanced technology such as AI that also have dual-use and cross-domain uses. In sum, the extent military AI poses risks to strategic stability will largely depend on the pace and scope of this technology to facilitate new ways to improve the delivery of and defense against nuclear weapons and strategic non-nuclear weapons. Part III will investigate these dynamics.

While the potential strategic effects of AI are not unique or exclusive to this technology, the confluence of five interrelated trends, examined in the remainder of this book, weighs heavily on the pessimistic side of the instability–stability ledger: (1) the rapid technological advancements and diffusion of military AI;[65] (2) the inherently destabilizing characteristics of AI technology; (3) the multi-faceted possible intersections of AI with nuclear weapons; (4) the interplay of these intersections with strategic non-nuclear capabilities; and finally (5) the backdrop of a competitive multipolar nuclear world order, which may entice states to prematurely deploy unverified, unreliable, and unsafe AI-augmented weapons into combat situations.

## Conclusion

This chapter conceptualized AI and modern technological change in the context of nuclear weapons and strategic stability. The first section found that strategic stability is ultimately a product of a complex interplay of myriad forces, and technology is best viewed as a variable that can have both stabilizing and destabilizing effects. Technology has long been used to augment, automate, and enhance human behavior and decision-making in a military context, and so far, the 'human factor' has had a greater influence than technology on strategic stability.

The chapter found that the instability which emerged during the Cold War has important implications for the discourse on AI and strategic stability. First, the destabilizing impact of weapons proliferation becomes more acute in situations where technology has the potential to create asymmetric military advantages between rivals. Second, where the risk exists (or it is perceived) that a conventional conflict might cross the nuclear threshold, the implications for crisis instability and military escalation can be severe. Third, any meaningful definition of strategic stability must include advanced strategic non-nuclear weapons that adversaries would likely use to destroy an enemy's nuclear capabilities. Finally, nuclear modernization programs that are primarily motivated by technologically advanced capabilities can trigger arms racing dynamics that worsen crisis stability.

The second section revisited the existing analytical framework on emerging technology and nuclear stability to conceptualize AI. It found that the history of emerging technology and revolution in military affairs counsels caution in extrapolating from current (or past) technological trends. Technologies rarely evolve in the way futurists predict, and many defense innovations have had countervailing or conditional effects that have ameliorated naysayer's dystopian predictions. Thus, even if AI ultimately has a transformative strategic impact, this will likely be far more non-linear, iterative, and prosaic for the foreseeable future.

Military AI can be conceptualized as a natural manifestation of an established trend in emerging technologies (i.e. co-mingling of nuclear and conventional capabilities, compression of the decision-making timeframe, and the speed of war), which might persuade states to adopt destabilizing nuclear postures. While the potential strategic effects of AI are not unique to this technology, the confluence of several trends suggests that its trajectory will be fundamentally destabilizing. What makes AI in a military context such a destabilizing prospect? If military AI is potentially so destabilizing, then why are great powers pursuing this technology with such vigor? To address these questions, the next chapter unpacks the core characteristics of military AI.

## Notes

1 There is no single, universally accepted definition of 'strategic stability,' of the factors that contribute to stability (or instability), or agreed-upon metrics to measure it. For example, Henry A. Kissinger, "Arms Control, Inspection and Surprise Attack," *Foreign Affairs*, 38, 4 (July 1960), pp. 557–575; Glenn H. Snyder, *Deterrence and Defense: Toward a Theory of National Security* (Princeton, NJ: Princeton University Press, 1961), pp. 107–108; and Herman Kahn, *On Escalation: Metaphors and Scenarios* (New York: Praeger Publishers, 1965).

2 Robert Jervis, *The Meaning of the Nuclear Revolution: Statecraft and the Prospect of Armageddon* (Ithaca, NY: Cornell University Press, 1989).

3 See Eric Schlosser, *Command and Control: Nuclear Weapons, the Damascus Accident, and the Illusion of Safety* (New York: Penguin, 2013); and Scott D. Sagan, *The Limits of Safety: Organizations, Accidents, and Nuclear Weapons* (Princeton, NJ: Princeton University Press, 1993). On nuclear terrorism, see Graham Allison, *Nuclear Terrorism: The Ultimate Preventable Catastrophe* (New York: Owl, 2004); John Mueller, *Atomic Obsession: Nuclear Alarmism from Hiroshima to Al-Qaeda* (New York: Oxford University Press, 2010); and Keir A. Lieber and Daryl G. Press, "Why States Won't Give Nuclear Weapons to Terrorists," *International Security*, 38, 1 (Summer 2013), pp. 80–104.

4 Michael Gerson, "The origins of strategic stability," in Colby Elbridge and Michael Gerson (eds), *Strategic Stability: Contending Interpretations* (Carlisle, PA: Army War College, 2013), pp. 1–46.
5 For an influential study on the promotion and analysis of nuclear arms control and strategic stability, see Thomas C. Schelling and Morton Halperin, *Strategy and Arms Control* (New York: Twentieth Century Fund, 1961).
6 For literature on the 'second nuclear age', see Paul Bracken, *The Second Nuclear Age: Strategy, Danger, and the New Power Politics* (New York: Times Books, 2012); Colin S. Gray, *The Second Nuclear Age* (Boulder, CO: Lynne Rienner, 1999); and Keith Payne, *Deterrence in the Second Nuclear Age* (Washington, DC: Georgetown University Press, 1996).
7 See Lawrence Freedman, *The Evolution of Nuclear Strategy* (Basingstoke: Palgrave Macmillan 2003); Michael Quinlan, *Thinking About Nuclear Weapons: Principles, Problems, Prospects* (Oxford: Oxford, University Press 2009).
8 Freedman, *The Evolution of Nuclear Strategy*.
9 The 2010 *Nuclear Posture Review* states that the goal of US nuclear strategy is to "strengthen deterrence of regional adversaries," such as North Korea, while "reinforcing strategic stability" with Russia and China. US Department of Defense, *Nuclear Posture Review* (Washington, DC: US Department of Defense, 2010), xi.
10 Thomas C. Schelling, *Arms and Influence* (New Haven, CT: Yale University Press, 1966), p. 234.
11 Gerson, "The Origins of Strategic Stability," p. 26.
12 Lehman F. Ronald, "Future technology and strategic stability," in Elbridge and Gerson (eds), *Strategic Stability*, p. 147.
13 'Automation' (as opposed to 'autonomy') has been integrated with early warning systems – radars, satellites, and infrared systems – for several decades, to identify potential targets and then cue them to human commanders.
14 The foundational work on misperceptions in world politics is Robert Jervis, *Perception and Misperception in International Politics* (Princeton, NJ: Princeton University Press, 1976). Also see Charles Duelfer and Stephen Dyson, "Chronic Misperception and International Conflict: the US–Iraq Experience," *International Security*, 36, 1 (Summer 2011), pp. 75–78.
15 'Strategic stability' and 'strategic instability' are not necessarily mutually exclusive states, especially in a nuclear multipolar system. Instead, the two states can be viewed as a 'stability continuum,' whereby situations can range from extremely stable to extremely unstable. Altmann Jürgen and Frank Sauer, "Autonomous Weapon Systems and Strategic Stability," *Survival*, 59, 5 (2017), pp. 117–142, p. 110.
16 Sources of strategic 'surprise' can include: (1) anticipating, detecting, and evaluating change – and reactions to these changes; (2) predicting counter-reactions and counter-trends; and (3) compensating for 'emergent behavior' that occurs when a combination results in behavior not characteristic of the sum of the components, or at least not in an obvious manner. See Ronald, "Future Technology and Strategic Stability," p. 147.

17 For example, even though new US capabilities such as ballistic missile defense and conventional prompt global strike could potentially blunt a Russian nuclear strike, they would be unlikely to eliminate the threat. However, this technical reality has done little to diminish Russian efforts to develop comparable precision weapons as a hedge against the perceived threat posed by US prompt global strike to Russia's nuclear deterrence. See Vladimir V. Putin, "Being Strong: National Security Guarantees for Russia," *Rossiiskaya Gazeta*, February 20, 2012, available from archive.premier.gov.ru/eng/events/news/18185/ (accessed February 20, 2021).

18 Political psychology literature demonstrates that the propensity for actors to live in different perceptual realities is not limited to relations between adversaries. Allies, too, can misperceive one another, and thus fail to understand (or recognize) others' perceptions of them. For example, see Richard Neustadt, *Alliance Politics* (New York: Columbia University Press, 1970).

19 Stephen J. Cimbala, *The Dead Volcano: The Background and Effects of Nuclear War Complacency* (Westport, Connecticut: Praeger, 2002), p. 66.

20 Prospect theory demonstrates that fear (especially the risk of suffering losses) is usually a more powerful motivator for the conflict and crisis than aggression or the desire for expansion. Robert Jervis *How Statesmen Think: The Psychology of International Politics* (Princeton, NJ: Princeton University Press, 2017), chapter 4.

21 Ibid., pp. 218–219.

22 The binary distinction between deliberate and unintentional use of nuclear weapons can be problematic. 'Unintentional' uses of nuclear weapons could be deliberate: smaller numbers of warheads could be used to create the perception – as part of a broader deception stratagem – that a deliberate use was unintentional, i.e. accidental or unauthorized. Also, the deliberate use of nuclear weapons based on a false (or manipulated) assessment – or in response to a false alarm – might blur the lines of intentionality. For example, a recent study demonstrated that different ranges of categories of escalation – i.e. unauthorized, unintended, and deliberate use – were all based on mistaken assumptions. Sico van der Meer, "Reducing Nuclear Weapons Risks: A Menu of 11 Policy Options," *Policy Brief*, Clingendael Netherlands Institute of International Relations, June 2018, www.clingendael.org/sites/default/files/2018-06/PB_Reducing_nuclear_weapons_risks.pdf (accessed February 20, 2021).

23 The risk of unintended escalation is challenging to quantify with any degree of confidence or accuracy; irrespective of the actual impact on stability actions by one side (or both) to manipulate risk will likely be perceived by the other as destabilizing. Further, the fear of escalation (or risk intolerance) might not be tied to a specific (or singular) destabilizing force or capability. Aaron R. Miles, "The dynamics of strategic stability and instability," *Comparative Strategy*, 35, 5 (2016), pp. 423–437, p. 429 and 437.

24 'Crisis stability' is only relevant in situations where states have nuclear warfighting in place, or a proclivity towards the use of nuclear weapons for pre-emptive strikes – as the US, Russia, and China currently have. For Russian nuclear

doctrine see Matthew Rojansky "Russia and strategic stability," in Elbridge and Gerson (eds), *Strategic Stability*, pp. 295–342. For China, see Benjamin Zala and Andrew Futter, "Coordinating the Arm Swing with the Pivot: Nuclear Deterrence, Stability and the US Strategy in the Asia-Pacific," *The Pacific Review*, 28, 3 (2015), pp. 367–390; and James S. Johnson, "Chinese Evolving Approaches to Nuclear 'Warfighting': An Emerging Intense US–China Security Dilemma and Threats to Crisis Stability in the Asia Pacific," *Asian Security*, 15, 3 (2019), pp. 215–232. For a recent analysis of the US thinking, see Charles L. Glaser and Steve Fetter, "Should the United States Reject MAD? Damage Limitation and US Nuclear Strategy Toward China," *International Security*, 41, 1 (Summer 2016), pp. 49–98.
25 Thomas C. Schelling, *The Strategy of Conflict* (Cambridge, MA: Harvard University Press, 1960).
26 Jervis, *How Statesmen Think*, p. 222.
27 Similar to 'strategic stability,' the concept of 'arms race' is also contestable. An 'arms race' is defined by adversaries' strenuous efforts to outmatch one another in the accumulation of weapons and is designed to shift the balance of military power decisively. Barry Buzan and Eric Herring, *The Arms Dynamic in World Politics* (Boulder, CO and London: Lynne Reinner), p. 77; and *Idem*, "Arms races and other pathetic fallacies: a case for deconstruction," review essay on *Plowshares into Swords: Arms Races in International Politics, 1840–1991*, Grant T. Hammond, *Review of International Studies*, 22 (July 3, 1996), pp. 323–336.
28 The question of whether peacetime, crisis, and wartime stability are necessary for strategic stability, or one (or two) is sufficient, lies beyond the scope of this study but merits further research.
29 Schelling and Halperin, *Strategy and Arms Control*.
30 Prospect theory can also explain how crisis instability can lead to all-out warfare even though neither side desires this outcome. See Jervis *How Statesmen Think*, p. 95.
31 Miles, "The dynamics of strategic stability and instability," pp. 423–437.
32 Psychological factors that may increase the risks of crisis instability include the following: (1) under the stress of a crisis (especially complex situations) people are subject to cognitive bias – or other constraints that shade their thinking – that can impact the quality of decision-making, without them necessarily being aware of this; (2) psychological processes can cause people to exaggerate the likelihood of an imminent attack during a crisis and, simultaneously, overstate the strategic value of striking first; (3) people often do not have fixed (or even stable) preferences and have a poor intuitive grasp of probability; and (4) humans often underestimate the extent that others perceive their actions as threatening. These factors raise questions about the assumption that, during a crisis, commanders can depend on a common sense of rationality with the other side. See Jervis, *How Statesmen Think*, chapter 10; and B.A. Thayer, "Thinking About Nuclear Deterrence Theory: Why Evolutionary Psychology Undermines its Rational Actor Assumptions," *Comparative Strategy*, 26, 4 (2007), pp. 311–323.

33 Bargaining situations in the context of nuclear weapons involve frequently competing for incentives, including: (1) deterrence: dissuading an adversary from doing something they want to do; (2) compellence: forcing an adversary to do something they do not wish to do; (3) assurance: convincing allies that security guarantees are credible; and (4) reassurance: convincing an adversary it will not be attacked so long as they refrain from provocative behavior. See Jervis, *Perception and Misperception in International Politics*, chapter 3; and Andrew H. Kydd and Roseanne W. McManus, "Threats and Assurances in Crisis Bargaining," *Journal of Conflict Resolution*, 61, 2 (2017), pp. 68–90.

34 Benjamin Zala, "How the Next Nuclear Arms Race will be Different from the Last One," *Bulletin of the Atomic Scientists*, 75, 1 (2019), pp. 36–43; and Colin S. Gray and Dale C. Walton, "The geopolitics of strategic stability," in Elbridgeand Gerson (eds), *Strategic Stability*, p. 93.

35 Gray and Walton, "The Geopolitics of Strategic Stability," p. 110.

36 Winner Langdon, *Autonomous Technology: Technics-out-of-Control as a Theme in Political Thought* (Cambridge, MA: MIT Press, 1977).

37 James M. Acton, "Reclaiming strategic stability," in Elbridge and Gerson (eds), *Strategic Stability*, pp. 138–139.

38 Jervis, *The Meaning of the Nuclear Revolution*, p. 45. In the theory of the nuclear revolution, see Kenneth N. Waltz, "Nuclear Myths and Political Realities," *American Political Science Review*, 84, 3 (September 1990), pp. 731–745.

39 Robert Jervis, *The Illogic of American Nuclear Strategy* (Ithaca, NY: Cornell University Press, 1984); Waltz, "Nuclear Myths and Political Realities;" Charles L. Glaser, *Analyzing Strategic Nuclear Policy* (Princeton, NJ: Princeton University Press, 1990); and Van Evera, *Causes of War*.

40 Keir A. Lieber and Daryl G. Press, "The New Era of Counterforce: Technological Change and the Future of Nuclear Deterrence," *International Security*, 41, 4 (2017), pp. 9–49.

41 Ronald, "Future technology and strategic stability," in Elbridge and Gerson (eds), pp. 147–199.

42 In this view, a combination of pre-emptive counterforce technology and missile defense capabilities that blunted an adversary's nuclear deterrent (or second-strike capacity) would be highly destabilizing. Lieber and Press, "The New Era of Counterforce."

43 Thomas Schelling argued that risk manipulation and *uncertainty* play critical roles in deterrence. Thomas C. Schelling, *Arms and Influence*, pp. 92–125.

44 Bargaining situations are those where actors gain from cooperating (as opposed to zero-sum, where various possible outcomes favor one side more than the other). See James D. Fearon, "Rationalist Explanations for War," *International Organization*, 49, 3 (1995), pp. 379–414; and Erik Gartzke, "War is in the Error term," *International Organization*, 53, 3 (1999), pp. 367–587.

45 Four trends in military technology and doctrine will likely exacerbate the future threats posed by 'entanglement:' (1) Technological advances in conventional precision missiles and cyber-weapons; (2) increasing vulnerability of nuclear and conventional command, control, communications, and intelligence (C3I)

systems to cyber-attacks; (3) growing military reliance on dual-use C3I capabilities; and (4) the recent emergence of conventional warfighting military doctrines (China, Russia, and the US), which explicitly envisage (early and pre-emptive) attacks on an adversary's C3I dual-use assets. James M. Acton, "Escalation Through Entanglement: How the Vulnerability of Command-and-Control Systems Raises the Risks of an Inadvertent Nuclear War", *International Security*, 43, 1 (Summer 2018), pp. 56–99.
46 Schelling and Halperin, *Strategy and Arms Control*. p. 37.
47 Austin Long and Brendan Rittenhouse Green, "Stalking the Secure Second Strike: Intelligence, Counterforce, and Nuclear Strategy," *Journal of Strategic Studies*, 38, 1–2 (2015), pp. 38–73; Owen R. Coté Jr., *The Third Battle: Innovation in the US Navy's Silent Cold War Struggle with Soviet Submarines* (Newport, RI: Naval War College, 2003); Peter Sasgen, *Stalking the Red Bear: The True Story of a US Cold War Submarine's Covert Operations Against the Soviet Union* (New York: St. Martin's Press, 2009).
48 Altmann Jürgen and Frank Sauer, "Autonomous Weapon Systems and Strategic Stability," *Survival*, 59, 5 (2017), pp. 117–142.
49 For definitions of the concepts of escalation threshold, escalation ladder, and escalation path see Herman Kahn, *On Escalation: Metaphors and Scenarios* (New York: Praeger, 1965), p. 37.
50 US Congress, Office of Technology Assessment, Ballistic Missile Defense Technologies, OTA-ISC-254 Washington DC: US Government Printing Office, September 1985), pp. 119 and 128.
51 James M. Acton, *Silver Bullet? Asking the Right Questions About Conventional Prompt Global Strike* (Washington, DC: Carnegie Endowment for International Peace, 2013).
52 Examples of 'strategic capabilities' include: long-range nuclear and conventional munitions (e.g. ICBMs), long-range penetrating bombers, shorter-range tactical (or theater) weapons that are or can be forward deployed. Offensive cyber and counter-space (i.e. anti-satellite) weapons have also emerged as strategic capabilities. Finally, defense systems (e.g. ballistic missile defense systems) can also be viewed as strategic in as much they are intended (or able) to impair a state's ability to respond at a strategic level.
53 The worlds' official nuclear states – the US, Russia, China, India, Britain, and France – all eschewed recent calls by Washington for a 'world without nuclear weapons' and instead have embarked on nuclear modernization programs. To date, the US has lagged in this effort, and the detail of its modernization program remains inconclusive. Zala, "How the Next Nuclear Arms Race Will be Different From the Last One," pp. 36–43.
54 Lehman F. Ronald, "Future Technology and Strategic Stability," pp. 147–199.
55 Stephen D. Biddle, "The PAST as Prologue: Assessing Theories of Future Warfare," *Security Studies*, 8, 1 (1998), pp. 1–74.
56 See Steven Metz, *Armed Conflict in the 21st Century: The Information Revolution and Post-Modern Warfare* (Carlisle: Strategic Studies Institute, 2000); and Steven Metz and James Kievit, *Strategy and the Revolution in*

*Military Affairs: From Theory to Policy* (Carlisle: Strategic Studies Institute, 1995); and Stephen Biddle, *Military Power: Explaining Victory and Defeat in Modern Battle* (Princeton, NJ: Princeton University Press, 2004).

57 Bernard Brodie and Fawn M. Brodie, *From Crossbow to H-Bomb* (Bloomington: Indiana University Press, 1973), chapter 23.

58 Stephen P. Rosen, "The Impact of the Office of Net Assessment on the American Military in the Matter of the Revolution in Military Affairs," *Journal of Strategic Studies*, 33, 4 (2010), pp. 469–482.

59 Andrew Liptak, "The US Air Force Has a New Weapon Called THOR That Can Take Out Swarms of Drones," *theverge*, June 21, 2019, www.theverge.com/2019/6/21/18701267/us-air-force-thor-new-weapon-drone-swarms (accessed October 12, 2019).

60 Since the 1300s, only five revolutions in military affairs have occurred, and technological changes primarily drove only one to the tools of war (i.e. nuclear weapons). See MacGregor Knox and Williamson Murray (eds), *The Dynamics of Military Revolution, 1300–2050* (Cambridge: Cambridge University Press, 2001).

61 Fu Wanjuan, Yang Wenzhe, and Xu Chunlei "Intelligent Warfare, Where Does it Not Change?" *PLA Daily*, January 14, 2020, www.81.cn/jfjbmap/content/2020-01/14/content_252163.htm (accessed January 15, 2020).

62 Bernard Brodie, "Technological change, strategic doctrine, and political outcomes in historical dimensions of national security problems," in Klaus Knorr (ed.), *Historical Dimensions of National Security Problems* (Lawrence, KS: University Press of Kansas, 1976), p. 263.

63 Lieber and Press, "The New Era of Counterforce," pp. 9–49.

64 H.M. Kristensen, M. McKinzie, and T.A. Postol, "How US Nuclear Force Modernization is Undermining Strategic Stability: The Burst-Height Compensating Super-Fuze," *Bulletin of the Atomic Scientists*, March 1, 2017, https://thebulletin.org/2017/03/how-us-nuclear-force-modernization-is-undermining-strategic-stability-the-burst-height-compensating-super-fuze/ (accessed May 1, 2019).

65 Where primarily commercial forces drive military innovation, technology tends to mature and diffuse at a faster pace compared to technologies with only a military utility, such as stealth technology. Michael C. Horowitz, *The Diffusion of Military Power: Causes and Consequences for International Politics* (Princeton, NJ: Princeton University Press, 2010).

# Part II

Military AI superpowers

# 3

# New challenges to military-techno
# *Pax Americana*

Why does the US view China's progress in dual-use AI as a threat to its first-mover advantage? How might the US respond to this perceived threat? This chapter considers the intensity of US–China strategic competition playing out within a broad range of AI and AI-enabling technologies (e.g. machine learning (ML), 5G networks, autonomy and robotics, quantum computing, and big-data analytics).[1] It describes how great-power competition is mounting within several dual-use high-tech fields, why these innovations are considered by Washington to be strategically vital, and how (and to what end) the US responds to the perceived challenge posed by China to its technological hegemony. The chapter uses the International Relations (IR) concept of 'polarity' (the nature and distribution of power within the international system) as a lens to view the shifting great-power dynamics in AI-related strategic technology (e.g. microchips, semiconductors, big-data analytics, and 5G data transmission networks).[2]

The chapter argues that the strategic competition playing out within a broad range of dual-use AI and AI-enabling technologies will likely *narrow* the technological gap separating great military powers (notably the US and China) and, to a lesser extent, other technically advanced small–medium powers.[3] The chapter builds on the growing body of literature that reinforces the perception in the US that China's pursuit of AI technologies will threaten the first-mover advantage that the US has in a range of dual-use (and military-specific) AI applications.[4] Because of this perceived threat, Washington will likely consider even incremental progress by China through a military lens, and thus treat any progress as a national security threat. However, certain areas of Chinese progress will likely be considered by the US as more strategically germane than others. Chinese AI-enabled predictive algorithms accurately capture the diverse variables and complex interactions of modern warfare, for example, that might offer China a full-spectrum multi-domain command and control capability, which could be viewed by the US as a major strategic threat (see chapter 8).

What are the implications of US–China defense innovation for the strategic balance and stability and, in particular, for efforts by the US to sustain its first-mover advantages (i.e. locking in a period of military advantages as a result of early technological innovation) in advanced military technology? Will the increasingly competitive US–China relationship dominate world politics creating a new bipolar world order, as opposed to a multipolar one?[5] This chapter is an attempt to acquire greater insight into these questions, to understand better the shifting power dynamics and strategic competition in the development of AI-related and enabling technology, and the implications of these trends for strategic relations between great powers.

The chapter proceeds in four sections. The first section summarises the responses by US decision-makers and analysts to the debate about US decline and the rise of the narrative of an imminent shift to a multipolar order. This grand strategic overview will be contextualized with particular reference to the relative decline of the US vis-à-vis China and the implications of the US being displaced as a global hegemon. The second section then sets up the debate over rapid advances and proliferation of AI-related technology capabilities, through an exploration of those that view harnessing these capabilities as a central aspect of efforts to maintain Washington's unipolar dominance.

The third section examines the perception of the rise of a bipolar order divided between Washington and Beijing. Again using the lens of defense innovation, it analyzes the credibility of the popular idea that the US has been caught off guard by China's accomplishments in the development of AI-related technologies, and that as a result, the US risks losing its first-mover advantages in the adoption of AI on the future battlefield. The final section considers the nature of this particular 'arms race' in the context of a predicted shift toward a multipolar order,[6] in particular, the commercial driving forces and dual-use features of this dynamic, which *prima facie* intonate a much more diffused and multipolar reality – as opposed to a bipolar one.

This study uses a wide range of open-source Chinese language reports, in combination with commercial and defense-centric think-tank and research group reports, to benchmark both China's AI approach and US perceptions of these developments – ostensibly through a national security and military lens.[7]

## New challenges to US technological hegemony

In the post-Cold War era, a preoccupation of US policy-makers and analysts has been the nature and implications of US unipolarity. This discourse has centered on two interrelated questions. How long will unipolarity last? And is the pursuit of hegemony a viable or worthwhile strategic objective for the US to pursue? The preservation of the US liberal hegemonic role as unipole has been the overarching grand strategic goal of every post-Cold War administration from George H.W. Bush to Barack Obama.[8] Having outlined the prominent strands and voices about how the US should (and can) respond to the notion of decline and the rise of a multipolar order, the analysis that follows uses the polarity in IR as a lens to explore how the US is positioning itself vis-à-vis China in AI technology – preparing for bipolarity with China or reluctantly accepting multipolarity?[9]

As the opening chapter showed, world leaders quickly recognize the transformative potential of AI as a critical component of national security.[10] This is largely driven by the perceived challenges posed by rising revisionist and dissatisfied powers (notably China and Russia).[11] The US Department of Defense (DoD) released, in 2016, a 'National Artificial Intelligence Research and Development Strategic Plan' – one of a series of studies about AI ML – on the potential for AI to reinvigorate US military dominance.[12] In the context of managing the potential flashpoints in the Taiwan Straits, the South China Seas, and Ukraine, former US Secretary of Defense Ashton Carter opined that China and Russia are the US's "most stressing competitors." Carter added that China and Russia continue to "advance military systems that seek to threaten our [US] advantages in specific areas" (including AI) and in "ways of war that seek to achieve their objectives rapidly, before, they hope, we [the US] can respond."[13]

To capitalize on the US's comparative advantage in private sector innovation, and to circumvent the burdensome military industrial-acquisition process, the DoD also established the Defense Innovation Unit Experimental to foster – albeit with mixed success – closer collaboration between the Pentagon and Silicon Valley.[14] In a similar vein, the recent summary of the DoD's debut AI strategy, *Artificial Intelligence Strategy*, stated that "China and Russia are making significant investments in AI for military purposes" that *"threaten to erode* our [US] technological and operational advantages" (emphasis added).[15] In response, the US must "adopt [military-use] AI to *maintain its strategic position*, prevail on future battlefields, and safeguard this [i.e. US-led] order" (emphasis added).[16]

Recent IR scholarship demonstrates that rising powers' pursuit of defense innovation is most likely to raise security concerns for the dominant state when the behavior of the rising power is viewed as an attempt to undermine the existing order – rules, norms, and governing institutions[17] – particularly if the new order conflicts with the dominant states' national security interests. China and Russia have developed a range of military-use

AI technologies as part of a broader strategic effort to simultaneously exploit perceived US military vulnerabilities and reduce their vulnerabilities.[18]

In a quest to become a 'science and technology superpower,' and catalyzed by AlphaGo's victory (or China's 'Sputnik moment'),[19] Beijing launched a national-level AI-innovation agenda for 'civil-military fusion' – or US Defense Advanced Research Projects Agency (DARPA) with Chinese characteristics. More on China's AI agenda below.[20] Similarly, the Russian private sector has benefited from state-directed support of human capital development and early investment in advanced technologies, in a broader effort to substitute its continued dependence upon Western technology with indigenous technologies, despite Russia's weak start-up culture. In short, national-level objectives and initiatives demonstrate recognition by great military powers of the potential military-technological transformative potential of AI for national security and for strategic stability between great military powers (see chapter 4).

US analysts and policy-makers have suggested a range of possible responses to these emerging security threats to preserve US technological leadership, which harnesses US natural advantages to push back against the rising great military powers in the multipolar order.[21] First, the DoD should fund and lead AI-simulated war games and red-teaming creative thinking exercises, to investigate existing and new security scenarios involving disruptive AI innovations. Second, the US needs to leverage its world-class think-tank community, academics, AI experts, computer scientists, and strategic thinkers to assess the implications of AI for a range of security scenarios and devise a long-term AI strategic agenda to meet these challenges.

Third, the US should prioritize DoD AI-based research and development (R&D) to leverage the potential low-cost, multiplier advantages of AI technologies (i.e. autonomy and robotics), and mitigate potential vulnerabilities and risks. Fourth, the US defense community should actively invest in and establish a commanding position in the nascent development of 'counter-AI' capabilities – both offensive and defensive.

Fifth, the US defense community (e.g. DARPA, the US Intelligence Advanced Research Projects Activity, Defense Innovation Board; the Office of Naval Research; and the National Science Foundation) should request more funds for AI-related research to combat the competition for talent and information in AI, and actively support university programs to ensure the US retains its relative talent pool advantages – in particular, vis-à-vis China. Finally, the Pentagon should fund additional R&D in reliable fail-safe and safety technology for AI-systems – especially military AI applications and tools.

## Washington's new Sputnik moment?

As AI military applications have grown in scale, sophistication, and lethality (see chapter 1), many in the US defense community have become increasingly alarmed about the implications of this trend for international competition and national security.[22] In his opening comments at 'The Dawn of AI' hearing, US Senator Ted Cruz stated, "ceding leadership in developing AI to China, Russia, and other foreign governments will not only place the United States at a technological disadvantage, but it could have *grave implications for national security.*"[23] Similarly, Director of US National Intelligence Daniel Coats recently opined, "the implications of our adversaries' abilities to use AI are *potentially profound and broad.*"[24]

Given the anticipated national security value China and Russia, in particular, attach to dual-use (i.e. military and civilian) AI-related technologies – notably autonomy and robotics, quantum communications, and 5G networks discussed below – several defense analysts have characterized the inexorable pace and magnitude of emerging security technologies (especially AI) as a 'Sputnik moment.' This in turn might portend a new military revolution (or be perceived as such), triggering a global AI arms race and changing the character (and even the nature) of warfare.[25] As chapters 1 and 2 showed, emerging security technology is only one facet of a broader trend toward increasing the speed of modern (conventional and nuclear) war, and shortening the decision-making timeframe associated with advances in weapon systems, such as cyber capabilities, direct-energy weapons, quantum communications, anti-satellite weapons, and hypersonic technology. The coalescence of these trends may lead to arms-race instability between great military powers, as rival states modernize their capabilities to reduce their perceived vulnerabilities.

Exponentially accelerated military-technological competition – in research, adoption, and deployment – of AI-related subset technologies (i.e. 5G networks, the internet of things (IoT), robotics and autonomy, additive manufacturing, and quantum computing) does not *necessarily* mean an 'arms race' is taking place. Instead, framing great-power strategic competition (especially US–China) in this way risks the adoption of operational concepts and doctrine, which in turn increases the likelihood of arms racing spirals occurring.[26] According to the DoD's Joint Artificial Intelligence Center (JAIC) former Director Lt. General Jack Shanahan, "it's strategic competition, *not an arms race*. They're [China] going to keep doing what we're doing; we [the US] acknowledge that."[27] Shanahan added: "What I don't want to see is a future where our [US] potential adversaries [China] have a fully AI-enabled force, and we do not."[28]

In response to a growing sense of consternation within the US defense community, the Pentagon has authored several AI-related programs and initiatives designed to protect US superiority on the future digitized battlefield (e.g. the Third Offset, Project Maven, DARPA's 'AI Next Campaign,' the establishment of the JAIC, the cloud-enabled Joint Common Foundation (JCF) platform, and the Pentagon's *AI Strategy*).[29] Taken together, these initiatives demonstrate the perceived gravity of the threat posed to US national security from the pursuit of AI-related capabilities by near-peer states (i.e. China and Russia) – notably autonomy and robotic systems, AI-augmented cyber offense, predictive decision-making tools, and enhancements to ISR capabilities, discussed in Part III of the book – to enhance their military power asymmetrically vis-à-vis the US.

For example, in response to Chinese strategic interest in AI, the Defense Innovation Unit proposed greater scrutiny and restrictions on Chinese investment in Silicon Valley companies.[30] This behavior typifies a broader concern that synergies created by China's civil-military fusion strategy could allow the technology, expertise, and intellectual property shared between US and Chinese commercial entities to be transferred to the People's Liberation Army (PLA).[31]

Moreover, broader US national security concerns relating to Chinese efforts to catch up (and even surpass) the US in several critical AI-related enabling technologies has prompted Washington to take increasingly wide-ranging and draconian steps to counter this *perceived* national security threat.[32] Against the backdrop of deteriorating US–China relations, responses such as these could accelerate the decoupling of cooperative bilateral ties between these two poles; increasing the likelihood of strategic competition, mutual mistrust, and negative action–reaction dynamics known as a security dilemma – which continues to manifest in other military technologies, including cyber, missile defense, hypersonic weapons, and counter-space capabilities (see chapters 6 and 7).[33]

Washington's alarmist tone and policy responses to the perceived threat posed by China's bid for technological leadership reveals that when we compare the public narratives surrounding the 'new multipolarity' thesis with what is happening, two things emerge.[34] First, the nature of the emerging great-power competition in AI suggests that a shift to US–China bipolarity (rather than multipolarity) is more likely in the short–medium term. Second, even in the event that China surpasses the US in AI (which many experts consider a strong possibility), it still trails the US in several qualitative measures (for now at least), which coalesce to preserve Washington's lead in the development of AI-related technologies.[35]

The US has the world's largest intelligence and R&D budgets, world-leading technology brands, academic research and innovation (discussed later in the chapter), and the most advanced (offense and defensive) cyber capabilities. Whether these advantages will be enough for Washington to forestall a shift in the military balance of power in the event China catches up or leap-frogs the US in AI – either through mimicry, espionage, or indigenous innovation – and can convert these gains (at a lower cost and less effort than the US) into potentially game-changing national security capabilities is, however, an open question.

China is, by some margin, Washington's closest peer-competitor in AI-related technology. Beijing's 2017 New Generation AI Development Plan identified AI as a core "strategic technology" and a new focus of "international competition." China's official goal is to "seize the strategic initiative" (especially vis-à-vis the US) and achieve "world-leading levels" of AI investment by 2030 – targeting more than US$150 billion in government investment.[36] Beijing has leveraged lower entry barriers to collect, process, and disseminate data within China to assemble a vast database to train AI systems.

According to a recent industry report, China is on track to possess 20 percent of the world's share of data by 2020, and has the potential to have over 30 percent by 2030.[37] Further, these efforts could be enhanced by the synergy and diffusion of a range of disruptive technologies such as ML, quantum technology,[38] 5G networks, and electromagnetics. In addition to the availability of vast data-sets, comparing the AI-capabilities of the US and China also incorporates wider qualitative and quantitative measures such as hardware, high-quality ML algorithms, private-public sector collaboration and broader technological, and scientific initiatives and policies.[39]

State-directed Chinese investment in the US AI market (in particular early-stage innovations) has also become increasingly active; in several instances, Chinese investment has competed directly with the DoD.[40] In 2017, for example, a Chinese state-run company Haiyin Capital outmaneuvered the US Air Force's efforts to acquire AI software developed by Neurala.[41] Incidences such as these are indicative of broader US concerns relating to China's proclivity for industrial espionage in its race to catch up (and overtake) the US in several strategically significant AI-related dual-use fields (e.g. semiconductors, robotics, autonomous vehicles, 5G networks, cyber, the IoT, big-data analytics, and quantum communications).[42] Industrial espionage, however, can only take the Chinese so far. The development of China's national innovation base, expertise, and capacity – even if that foundation builds on industrial and other types of espionage and mimicry – is a broader trend of which the DoD also appears to be cognizant.[43]

Among these critical enabling technologies that could fundamentally change the future of warfare are next-generation data transmission networks. The strategic importance of 5G networks as a critical future military technological enabler was demonstrated during the protracted and ongoing tensions between China's Huawei and Washington.[44] Experts view 5G as a cornerstone technology to increase the speed, volume, and stability dataloads, to reduce the latency (i.e. accelerate network response times), and to enhance mobile digital communications and information-sharing capabilities in both military and commercial contexts. According to an AI and telecommunications researcher at the University of Electronic Science and Technology of China, "the 5G network and the IoT enlarge and deepen the cognition of situations in the battlefield by several orders of magnitude and *produce gigantic amounts of data, requiring AI to analyze and even issue commands*"[45] (emphasis added).

Against the backdrop of rising tensions in the US–China relationship on a plethora of interconnected policy arenas (e.g. trade, and geopolitical influence in the Asia-Pacific), the technological race for access and control of critical enablers that will connect sensors, robotics, autonomous weapon systems, and the exchange of vast volumes of data in real-time through AI ML techniques on the digitized battlefield, will become increasingly intense and strategically motivated.[46]

In 2017, Chinese President Xi Jinping explicitly called for the acceleration of the military's intelligentization agenda to prepare China for future warfare against a near-peer adversary, namely the US.[47] Although Chinese think-tanks and academic discourse are generally poor at disseminating their debates and content, open-source evidence suggests a clear link between China's political agenda related to the 'digital revolution,' Chinese sovereignty and national security, and the current public debate surrounding the rejuvenation of the Chinese nation as a great power.[48] In short, national security is ultimately interpreted by China (and the US) as encompassing economic performance.[49]

President Xi's belt-and-road-initiative (BRI), and the virtual dimension the 'Digital Silk Road,' are high-level efforts designed to ensure that the mechanisms, coordination, and support for this agenda will become increasingly normalized.[50] Xi recently stated that AI, 'big data,' cloud storage, cyberspace, and quantum communications, were amongst the "liveliest and most promising areas for civil-military fusion."[51] While BRI investment is predominantly in emerging markets with comparably low levels of technology maturity, human capital, and military power, the BRI framework supports a broader Chinese agenda to expand (or establish a new) geopolitical sphere of influence; to improve its position in the future distribution of power – especially vis-à-vis the US.[52]

Towards this end, in 2017, Beijing established the 'Military-Civil Fusion Development Commission,' designed to expedite the transfer of AI technology from the commercial research centers to the military.[53]

Recent Chinese achievements in AI demonstrate Beijing's potential to realize this goal. For example, in 2015, Baidu reportedly designed AI software capable of surpassing human levels of language recognition, a year before Microsoft achieved a similar feat.[54] China is actively researching a range of air, land, and sea-based autonomous vehicles in the defense realm.[55] In 2017, following reports of a computer-simulated swarming destroying a missile launcher, a Chinese university with ties to the PLA demonstrated an AI-enabled swarming of 1,000 unmanned aerial vehicles at an airshow.[56] Also, open sources indicate that China is also pursuing a range of AI-enabled applications to augments its existing cyber (offensive and defensive) capabilities.[57]

In the case of quantum technology, the potential convergence between AI and quantum computing could create promising synergies that Beijing intends to leverage to ensure it is at the forefront of the so-called 'quantum AI revolution.'[58] Moreover, Chinese analysts and strategists anticipate that quantum technologies will radically transform future warfare, with a strategic significance equal to nuclear weapons.[59] In 2015, Chinese researchers reportedly achieved a breakthrough in the development of a quantum ML algorithm, which may remove several technical bottle-necks (e.g. quantum radar, sensing, imaging, metrology, and navigation). This would allow greater independence from space-based systems – where China currently lags behind the US – enhancing ISR capabilities, potentially creating new vulnerabilities in US space-based GPS and stealth technology in future conflict scenarios – especially in the undersea domain.[60]

The evidence suggests a strong link between Beijing's pursuit of AI technology and its broader geopolitical objectives. This link has, in turn, reinforced the narrative within the US defense community, that China believes this technological transformation is an opportunity to strengthen its claim on the leadership – and eventual dominance – of the emerging technological revolution, having missed out on previous waves. In sum, despite the apparent economic issues at stake (i.e. the rents to be captured in the data-driven economy), the threat to US technological hegemony is generally interpreted through a military and geopolitical lens.[61]

By contrast, the increasingly strained relationship between the previous Trump administration and Silicon Valley will likely pose additional challenges to this critical partnership in the development of AI technologies for the US military.[62] Following a high-profile backlash from employees at Google, for example, the company announced in 2018 that it would discontinue its work with the Pentagon on Project Maven – a Pentagon program to build an AI-powered surveillance platform for unmanned aerial vehicles.[63] Several defense analysts and US government reports have noted the growing gap between the rhetoric and the research momentum (especially in AI and robotics), and the paucity of resources available, to make the US military more networked and integrated.[64]

Specifically, analysts highlight various shortcomings in the US defense innovation ecosystem, including inadequate funding to sustain long-term R&D, institutional stove-piping, and an insufficient talent pool to attract and retain top scientists in AI-related fields.[65] In its 2018 *AI Strategy*, the DoD committed to "consult with leaders from across academia, private industry, and the international community" and "invest in the research and development of AI systems."[66] Details of the implementation and funding arrangements for these broad principles remain mostly absent, however. Moreover, the apparent mismatch (even dissonance) between the rapid pace of commercial innovation in AI technologies, and the lagging timescales and assumptions that underpin the US DoD's existing procurement processes and practices could exacerbate these bilateral competitive pressures.[67]

China's pursuit of AI-enabling (especially dual-use) technologies will fuel the perception in Washington – accurate or otherwise –that China is intent on exploiting these strategically critical capabilities to fulfill China's broader revisionist goals.[68] Once the Digital Silk Road initiative reaches fruition, therefore, BRI might enable China's 5G, AI, and precision navigation systems to dominate the digital communications and intelligence of every nation within the BRI sphere of influence, as part of Beijing's broader strategic objective, to ensure the leadership of a new international order[69] or a separate China-led bounded order – China's version of the Greater East Asia Co-Prosperity sphere, or Halford Mackinder and Mahan's theories of world control.[70]

In addition to this unique scaling advantage, China's defense AI innovation has benefited from its approach to AI acquisition: a centralized management system where few barriers exist between commercial, academic, and national security decision-making. While many analysts consider China's centralized approach to the development of AI affords it unique advantages over the US, others posit that Beijing's AI strategy is far from perfect. Some analysts, for example, have characterized Beijing's funding management as inherently inefficient. These analysts note that China's state apparatus is inherently corrupt and that this approach tends to encourage overinvestment in particular projects favored by Beijing, which may exceed market demand.[71]

Besides, China is currently experiencing a shortage of experienced engineers and world-class researchers to develop AI algorithms. For instance, China has only thirty Chinese universities that produce indigenous experts and research products. As a corollary, industry experts have cautioned that Beijing's aggressive and centralized pursuit of AI could result in poorly conceptualized AI applications that adversely affect the safety of AI-enabled military applications, and increase the potential systemic risks associated with these innovations.[72] The comparatively measured pace of US military AI innovation might, therefore, in the longer term result in more capable tools, but without sacrificing safety for speed – even at the cost of falling behind China's AI quantitative lead in the short-term. The prioritization by the US of the development of robust, verifiable, and safe military AI technology might, however, come under intense pressure if China's progress in dual-use AI is perceived as an imminent threat to the US first-mover advantages.

## Arms racing dynamics in AI

As the most powerful nation-states and leaders in the development of AI, competitive tensions between China and the US have often evoked comparisons with the Cold War-era US–Soviet space race.[73] In response to the global AI arms race, and to sustain US superiority and first-mover advantages in AI, General John Allen and Spark Cognition CEO Amir Husain have argued that the US must push further and faster to avoid losing the lead to China (and to a lesser degree Russia) in the development of AI.[74]

While these depictions accurately reflect the *nature* of the increasingly intense competition in the development of AI-related technologies, the *character* of this particular arms race intimates a more multipolar reality – compared to the Cold war-era bipolar space race. Over time, this trend will likely elevate technically advanced small and middle powers (e.g. South Korea, Singapore, Israel, France, and Australia) to become pioneers in cutting-edge dual-use AI-related technology, and key influencers shaping the security, economics, and global norms and standards of these innovations in the future world order.[75]

While this broader trend will not necessarily run in lockstep with the US–China bipolar contest, it will inevitably be influenced and shaped by the ongoing competition between China and the US (and its allies) on setting global technological standards for AI, nonetheless. For these middle powers, the outcome of this contest and, particularly, how they position themselves on AI technology standards will have a significant impact on their ability to truly become cutting-edge AI innovators – independent of China and the US.

The commercial driving forces underlying emerging security technologies (i.e. hardware, software, and R&D), together with the inherently dual-use nature of these innovations, reduce the usefulness of the space race analogy.[76] In short, the particular problem-set associated with the Cold War-era bipolar structure of power (i.e. a deadly obsession with the other side's military capabilities) is, to date at least, far less intense in the context of contemporary competition in AI.[77] Where primarily commercial forces drive military innovation, and particularly when imitation is quicker and cheaper than innovation, technology tends to mature and diffuse at a faster pace compared to military-specific applications, such as stealth technology.[78] Second-mover catch-up possibilities in military-use AI through imitation are in an adversarial context; therefore, it will unlikely be a feasible avenue for states.

Will specific technologies diffuse more evenly than others? As the literature on the diffusion of military technology demonstrates: how states react to and assimilate innovations can have profound implications for strategic stability and, in turn, the likelihood of war.[79] In particular, the pace military actors diffuse technology can influence the relative advantages derived by states from being the first mover, which tends to be reversely correlated to the speed innovations are adopted.[80] As the costs and availability of computing power – an essential ingredient for AI ML systems – decrease, therefore, so technically advanced military powers will likely pull away from those actors who are more (or entirely) reliant on mimicry and espionage.[81] Moreover, this trend will likely be compounded if either the cost or complexity of AI algorithms increases, allowing AI first-movers to maximize their competitive advantages.[82]

The growing sense is that the proliferation of AI technologies driven by powerful commercial forces will inevitably accompany (and even accelerate) the shift toward multipolarity. However, important caveats need to accompany prognostications about the pace and nature of this transformation: the risks associated with the proliferation and diffusion of dual-use AI technologies across multiple sectors and expanding knowledge bases are very different prospects compared to arms racing between great-power military rivals. Thus, the development of military AI applications based on military-centric R&D would make it much more difficult and costly for smaller (and especially less technically advanced) states to emulate and assimilate successfully.[83]

Moreover, military organizations, norms, and strategic cultural interests and traditions will also affect how security technology is assimilated by militaries, potentially altering the balance of military power.[84] As a result, the interplay of technology and military power will continue to be a complex outcome of human cognition, institutions, strategic cultures, judgment, and politics.[85] Ultimately, success in sustaining or capturing the first-mover advantages in AI will be determined by how militaries develop doctrine and strategy to seize on the potential comparative benefits afforded by AI-augmented capabilities on the battlefield – Part III of the book will investigate this theme.

The pace of military-use AI diffusion to small and medium powers (and non-state and third-party entities) will likely be constrained by three key features of this emerging phenomenon: hardware constraints (e.g. physical processors) and integrating increasingly sophisticated software and hardware with internal correctness;[86] the algorithmic complexity inherent to AI ML approaches; and the resources and know-how to deploy AI code effectively.[87] These features mean that militaries will need to invest vast amounts of capital and resources in a broad range of disciplines, inter alia, psychology, cognitive science, communication, human-computer interaction, computer-supporter workgroups, and sociology – thereby gaining experience through trial and error, to fuse AI with advanced weapon systems such as autonomous vehicles, hypersonic weapons, and missile defense (see chapter 6). In short, states will find it very difficult to develop and deploy military-use AI applications from technology derived from general ancillary dual-use applications alone.[88]

As a corollary, China's advantages from its commercial lead in the adoption of AI and data-set ecosystems will not necessarily be easily translated into special-purpose military AI applications. China's strengths in commercial-use AI (e.g. 5G networks, e-commerce, e-finance, facial recognition, and various consumer and mobile payment applications) will need to be combined with specialized R&D and dedicated hardware to unlock their potential dual-use military applications and augment advanced weapon systems. Without the requisite level of resources, know-how, data-sets, and technological infrastructure, these constraints could make it very difficult for a new entrant to develop and deploy modular AI with the same speed, power, and force as the US or China.[89]

For instance, China and the US are in close competition to develop the supercomputers needed to collect, process, and disseminate the vast amounts of data that traditional computers can handle. While the US possesses more powerful computers, China trumps the US in terms of the number of supercomputers.[90] Thus, military-led innovations could potentially concentrate and consolidate leadership in this nascent field amongst current military superpowers (i.e. China, the US, and to a lesser extent Russia), and revive the prospect of bipolar strategic competition.[91] For now, however, it remains unclear how specific AI applications might influence military power, or whether, and in what form these innovations will translate into operational concepts and doctrine.[92]

In sum, the degree to which AI alters the military balance of power will depend, in large part, on the speed of the diffusion of this technology within the military structures of the US and China; as a function of human innovation, political agendas, and strategic calculation and judgment, against the backdrop of a multipolar world – and nuclear – order, and heuristic decision-making (or the propensity for compensatory cognitive short-cuts) associated with decisions taken under compressed timeframes in uncertain and complex environments.[93] As the tectonic plates of the global political system continue to shift, the US and China are on opposite sides of the divide, and China's neighbors – particularly US allies and partners in the region – will need to pick a side and choose whether to work with or separately from the US – (its credibility, legitimacy and ability to impose its will are fast eroding) or side with China.

## Conclusion

This chapter examined the intensity of US–China strategic competition playing out within a broad range of AI-related and enabling technologies. It made the following central arguments. First, while disagreement exists on the likely pace, trajectory, and scope of AI defense innovations, a consensus is building within the US defense community intimating that the potential impact of AI-related technology on the future distribution of power and the military balance will likely be transformational, if not revolutionary. These assessments have, in large part, been framed in the context of the perceived challenges posed by revisionist and dissatisfied great military powers (China and Russia) to the current US-led international order – rules, norms, governing institutions – and military-technological hegemony.

Second, the rapid proliferation of AI-related military technology exists concomitant with a growing sense that Washington has dropped the ball in the development of these disruptive technologies. If the perception that the US's first-mover advantage in a range of dual-use enabling strategic technologies (i.e. semiconductors, 5G networks, and IoT's) from rising (especially nuclear-armed) military powers such as China becomes prevalent, the implications for international security and strategic stability could be severe (see chapter 4).

In response to a growing sense of urgency within the US defense community cognizant of this prospect, the Pentagon has authored several AI-related programs and initiatives designed to protect US dominance on the future digitized battlefield. Further, broader US national security concerns relating to Chinese efforts to catch up (and even surpass) the US in several critical AI-related enabling technologies has prompted Washington to take increasingly wide-ranging and draconian steps to counter this perceived national security threat.[94]

Third, and related to the previous finding, the development of AI evocations of the Cold War-era space race as an analogy does not accurately capture the nature of the evolving global AI phenomena. Instead, compared to the bipolar features of the US–Soviet struggle, this innovation arms race intimates more multipolar characteristics. Above all, the dual-use and commercial drivers of the advances in AI-related technology will likely narrow the technological gap separating great military powers (chiefly the US and China) and other technically advanced small–medium powers. These rising powers will become critical influencers in shaping future security, economics, and global norms in dual-use AI.

In the case of military-use AI applications, however, several coalescing features of this emerging phenomena (i.e. hardware constraints, ML algorithmic complexity, and the resources and know-how to deploy military-centric AI code), will likely constrain the proliferation and diffusion of AI with militaries' advanced weapon systems for the foreseeable future. In turn, these constraints could further concentrate and consolidate the leadership in the development of these critical technological enablers amongst the current AI military superpowers (i.e. China and the US), which could cement a bipolar balance of power and the prospect of resurgent bipolar strategic competition.[95]

At the present time, the US has an unassailable first-mover advantage in a range of AI applications with direct (and in some cases singular) relevance in a military context. However, as China approaches parity, and possibly surpasses the US in several AI-related (and dual-use) domains, the US will increasingly view future technological incremental progress in emerging technologies – and especially unexpected technological breakthroughs or surprises – through a national security lens. Thus, responses to these perceived threats will be shaped and informed by broader US–China geopolitical tensions.[96]

These concerns resonated in the 2018 US *Nuclear Posture Review*. This emphasized that geopolitical tensions and emerging technology in the nuclear domain could coalesce with unanticipated technological breakthroughs in "new and existing innovations to "change the nature of the threats faced by the US and the "capabilities needed to counter them."[97] In sum, against the backdrop of US–China geopolitical tensions, and irrespective of whether China's dual-use applications can be imminently converted into deployable military-use AI, US *perceptions* of this possibility will be enough to justify severe countermeasures. What will the implications be for deterrence, escalation, and crisis stability of this increasingly competitive geopolitical dyad? How might AI alter the strategic stability between great powers, and increase the potential for asymmetrical conflict? The chapter that follows addresses these questions.

## Notes

1 Chapter 1 described the line between core AI and 'AI-related' technology as a blurred one. Core AI technology include machine-learning (and deep learning and deep networks subset), modeling, automated language, image recognition, sensor imaging, voice assistants, and analysis support systems. AI also intersects with many other technologies (or AI-related and AI-enabling technologies), including autonomous vehicles, big-data analytics, 5G networks, supercomputers, smart vehicles, smart wearable devices, robotics, semiconductors, 3D printing, and the IoT.
2 This chapter is derived in part from an article published in *The Pacific Review*, October 21, 2019, copyright Taylor & Francis, available online: https://doi.org/10.1080/09512748.2019.1676299 (accessed March 10, 2020).
3 However, there is a significant difference between narrowing the gap in specific fields and gaining an overall lead across all the categories (i.e. talent, research, development, hardware, data, and adoption) in the emerging AI race. In a recent report published by the Center for Data Innovation: "Overall, the United States currently leads in AI, with China rapidly catching up, and the European Union behind both. The United States leads in four of the six categories of metrics this report examines (talent, research, development, and hardware) [while] ... China leads in two (adoption and data)" (Daniel Castro, *Who is winning the AI race: China, the EU, or the United States?* (Washington, DC: Center for Data Innovation, 2019), p. 2. Chinese open-source data also confirms that the US is ahead of China in these categories (China Institute for Science and Technology Policy at Tsinghua University, July 2018).
4 For example, see Vincent Boulanin (ed.), *The Impact of Artificial Intelligence on Strategic Stability and Nuclear Risk Vol. I Euro-Atlantic Perspectives* (Stockholm: SIPRI Publications, May 2019); Andrew W. Moore, "AI and National Security in 2017," Presentation at AI and Global Security Summit, Washington, DC, November 1, 2017; Daniel S. Hoadley and Nathan J. Lucas, *Artificial Intelligence and National Security* (Washington, DC: Congressional Research Service, 2017), https://fas.org/sgp/crs/natsec/R45178.pdf, p. 2 (accessed August 10, 2019); Greg Allen and Taniel Chan, *Artificial Intelligence and National Security* (Cambridge, MA: Belfer Center for Science and International Affairs, 2017).
5 Polarity analysis focuses on whether the inter-state order is dominated by one (a unipolar order), two (a bipolar order), or three or more (a multipolar order) centers of power. 'Multipolarity' implies that no single state is unambiguously in the lead (or polar) in the international order. In contrast to 'bipolarity,' that implies much less ambiguity in the stratification of power surrounding two poles. In addition to military power, economic capacity, demographics, 'soft power,' and the broader social dimensions of state influence have been associated with the shift towards a multipolar order. William C. Wohlforth, "Unipolarity, status competition, and great power war," in John G. Ikenberry, Michael Mastanduno, and William C. Wohlforth (eds), *International Relations Theory and the*

*Consequences of Unipolarity* (Cambridge: Cambridge University Press, 2011), pp. 33–65. For critiques on this contested concept see Harrison R. Wagner, *War and the State: The Theory of International Politics* (Ann Arbor, MI: University of Michigan Press, 2009); and Randall L. Schweller, "Entropy and the Trajectory of World Politics: Why Polarity has become Less Meaningful," *Cambridge Review of International Affairs*, 23, 1 (2010), pp. 145–163.

6 In this context an 'arms race' refers to a "progressive, competitive peacetime increase in armaments by two states or coalition of states resulting from conflicting purposes or mutual fears," Samuel P. Huntington, "Arms Races: Prerequisites and Results," *Public Policy*, 8 (1958), pp. 41–86, p. 43.

7 Open-source data sources include: ii Media, "2017 China Artificial Intelligence Industry Special Research Report," April 6, 2017, www.sohu.com/a/132360429_468646 (accessed December 12, 2018); Dominic Barton et al., "Artificial Intelligence: Implications for China" (discussion paper, McKinsey Global Institute, April 2017), www.mckinsey.com/~/media/McKinsey/Featured%20Insights/China/Artificial%20intelligence%20Implications%20for%20China/MGIArtificial-intelligence-implications-for-China.ashx (accessed May 10, 2019); China State Council, "State Council Notice on the New Generation Artificial Intelligence Development Plan," July 8, 2017, www.gov.cn/zhengce/content/2017-07/20/content_5211996.htm (accessed March 10, 2020); Jingwang Li, "2017 China–US AI Venture Capital State and Trends Research Report. IT Juzi and Tencent Institute," 2017 (Full report in Chinese) http://voice.itjuzi.com/?p=16960 (accessed December 12, 2018); New Intellectual Report, "China Academy of Engineering: Unveiling of Artificial Intelligence 2.0 Era," February 21, 2017, www.sohu.com/a/126825406_473283 (accessed December 12, 2018); China AI Development Report 2018, China Institute for Science and Technology Policy at Tsinghua University, July 2018 www.sppm.tsinghua.edu.cn/eWebEditor/UploadFile/China_AI_development_report_2018.pdf (accessed May 10, 2019); Council on Foreign Relations. "Beijing's AI Strategy: Old-School Central Planning with a Futuristic Twist," August 9, 2017, www.cfr.org/blog/beijings-ai-strategy-old-school-central-planningfuturistic-twist (accessed December 12, 2018); CB Insights, "Advanced Search: Industry & Geography, Company Attributes, Financing & Exit," https://app.cbinsights.com (accessed May 10, 2019).

8 Christopher Layne, "This Time It's Real: The End of Unipolarity and the '*Pax Americana*'," *International Studies Quarterly* 56, 1 (2012), pp. 202–213, p. 2.

9 The significant reactions (both official and non-official) to the notion of US decline and a shift into a multipolar order can be categorized into three broad strands of thought: the 'denialists,' 'accepters,' and 'resisters.'

10 Robert O. Work, *Remarks by Defense Deputy Secretary Robert Work at the CNAS Inaugural National Security Forum, Speech* (Washington, DC: CNAS, July 2015), www.defense.gov/Newsroom/Speeches/Speech/Article/634214/cnas-defense-forum/ (accessed March 10, 2020).

11 Office of the Secretary of Defense, *Annual Report to Congress: Military*

and *Security Developments Involving the People's Republic of China, 2019* (Washington, DC: US Department of Defense, 2019).
12 National Science and Technology Council, *The National Artificial Intelligence Research and Development Strategic Plan*, Executive Office of the President of the United States (Washington, DC, October 2016).
13 US Department of Defense, *Remarks by Secretary Carter on the Budget at the Economic Club of Washington*, February 2, 2016, www.defense.gov/Newsroom/Transcripts/Transcript/Article/648901/remarks-by-secretary-carter-on-the-budget-at-the-economic-club-of-washington-dc/ (accessed May 10, 2019).
14 Fred Kaplan, "The Pentagon's Innovation Experiment," *MIT Technology Review*, December 16, 2016, www.technologyreview.com/s/603084/the-pentagons-innovation-experiment/ (accessed March 10, 2020).
15 US Department of Defense, "Summary of the 2018 Department of Defense Artificial Intelligence Strategy," https://media.defense.gov/2019/Feb/12/2002088963/-1/-1/1/SUMMARY-OF-DOD-AI-STRATEGY.PDF (accessed August 10, 2019).
16 Ibid.
17 See Emily O. Goldman and Richard B. Andres, "Systemic Effects of Military Innovation and Diffusion," *Security Studies*, 8, 4 (1999), pp. 79–125.
18 For example, Russia's focus on AI-enabled land-based infantry robotic systems is considered by analysts to be based on the self-perception that its armored and heavy infantry brigades and divisions are conventionally weaker compared to NATO's. In addition to AI, both China and Russia have also developed other technologically advanced – and potentially disruptive – weapons such as cyber warfare tools, stealth and counter-stealth technologies; counter-space; missile defense; and guided precision munitions – Part III of the book will consider the interplay of these advanced weapon systems with AI.
19 This refers to the shock experienced by Washington when the Soviet Union beat the US into space by launching the world's first satellite, *Sputnik 1*, in 1957; China experienced a similar shock when Google's DeepMind AlphaGo AI defeated the world's number one Go player, Ke Jie. Sputnik shocked the US into making massive investments in science and technology to compete technologically with the Soviet Union; China responded to AlphaGo by rapidly increasing investment in the field of AI and AI-enabling technologies.
20 The State Council Information Office of the People's Republic of China, "State Council Notice on the Issuance of the New Generation AI Development Plan," July 20, 2017, www.gov.cn/zhengce/content/2017-07/20/content_5211996.htm (accessed March 10, 2020).
21 For example, see Hadley and Lucas, *Artificial Intelligence, and National Security*; Robert O. Work and Shawn W. Brimley, *20YY Preparing for War in the Robotic Age* (Washington, DC: Center for a New American Security, 2014), www.cnas.org/publications/reports/20yy-preparing-for-war-in-the-robotic-age (accessed December 12, 2018); Edward Geist and Andrew Lohn, *How Might Artificial Intelligence Affect the Risk of Nuclear War?* (Santa Monica, CA:

RAND Corporation, 2018); and White House, Executive Order on Maintaining American Leadership in Artificial Intelligence, February 11, 2019, www.whitehouse.gov/presidential-actions/executive-order-maintaining-american-leadership-artificial-intelligence/ (accessed May 10, 2019).
22 James Johnson, "Artificial Intelligence & Future Warfare: Implications for International Security," *Defense & Security Analysis*, 35, 2 (2019), pp. 147–169.
23 Ibid., p. 17.
24 Ibid.
25 Sputnik-like events are likely because potential US adversaries (i.e. China and Russia) consider the exploitation and diffusion of cutting-edge technology to provide asymmetrical responses to Western capabilities. See Robert A. Divine, *The Sputnik Challenge* (New York and Oxford: Oxford University Press, 1993), and Asif A. Siddiqi, *Sputnik and the Soviet Space Challenge* (Gainesville, FL: The University of Florida Press, 2000).
26 Recent hyperbolic and alarmist statements about an impending AI arms race are reminiscent of the misleading rhetoric surrounding the so-called 'cyber bombs' in the battle against ISIS. Heather M. Roff, "The Frame Problem: The AI 'Arms Race' Isn't One," *Bulletin of the Atomic Scientists* (2019), 75, 3, pp. 1–5.
27 Lt. General Jack Shanahan, Media Briefing on AI-Related Initiatives within the Department of Defense (Washington, DC: US Department of Defense August 30, 2019) www.defense.gov/Newsroom/Transcripts/Transcript/Article/1949362/lt-gen-jack-shanahan-media-briefing-on-ai-related-initiatives-within-the-depart/ (accessed July 10, 2019).
28 Ibid.
29 James Johnson, "The End of Military-Techno *Pax Americana*? Washington's Strategic Responses to Chinese AI-Enabled Military Technology," *The Pacific Review*, www.tandfonline.com/doi/abs/10.1080/09512748.2019.1676299?journalCode=rpre20 (accessed February 20, 2021).
30 Tom Simonite, "Defense Secretary James Mattis envies Silicon Valley's AI Ascent," *Wired.com*, November 8, 2017, www.wired.com/story/james-mattis-artificial-intelligence-diux/ (accessed July 10, 2019).
31 Carolyn Bartholomew and Dennis Shea, US–China Economic and Security Review Commission – *2017 Annual Report* (Washington, DC: The US–China Economic and Security Review Commission, 2017), p. 507; and Testimony of Jeff Ding before the US–China Economic and Security Review Commission, Hearing on Technology, Trade, and Military-Civil Fusion: China's Pursuit of Artificial Intelligence, New Materials and New Energy, June 7, 2019, www.uscc.gov/hearings/technology-trade-and-military-civil-fusion-chinas-pursuit-artificial-intelligence-newpdf (accessed July 1, 2019).
32 For example, Washington's ongoing efforts to stifle China's progress in 5G networks and semiconductors in the apparent defense of national security.
33 Robert Jervis, *Perception and Misperception in International Politics* (Princeton, NJ: Princeton University Press, 1976), chapter 3; and James Johnson, "Washington's Perceptions and Misperceptions of Beijing's Anti-Access Area-

Denial (A2–AD) 'Strategy': Implications for Military Escalation Control and Strategic Stability," *The Pacific Review*, 30, 3 (2017), pp. 271–288.

34 Benjamin Zala, "Polarity Analysis and Collective Perceptions of Power: The Need for a New Approach," *Journal of Global Security Studies*, 2, 1 (2017), pp. 2–17.

35 Kai-Fu Lee, *AI Superpowers: China, Silicon Valley, and the New World Order* (New York: Houghton Mifflin Harcourt, 2018).

36 Recent research suggests that China's AI investment targets are falling short of expectation. Some analysts predict the gulf between China and the US will continue to expand, with the US expected to reach a 70% market share in venture capital AI investment within the next 5–10 years. Deborah Petrara, "China's AI Ambition Gets a Reality Check as the USA Reclaims Top Spot in Global AI Investment," *Bloomberg Business Wire*, October 30, 2019, www.bloomberg.com/press-releases/2019-10-30/china-s-ai-ambition-gets-a-reality-check-as-the-usa-reclaims-top-spot-in-global-ai-investment (accessed November 10, 2019).

37 Will Knight, "China's AI Awakening," *MIT Technology Review*, October 10, 2017, www.technologyreview.com/s/609038/chinas-ai-awakening/ (accessed July 10, 2019).

38 The field of quantum ML is still in its infancy; to date, few ideas have been proposed regarding methods to utilize – even at a theoretical level – quantum computing to speed up ML and inference. In October 2019, for example, Google announced that it had reached a significant milestone in demonstrating quantum viability, the significance of the achievement has been disputed. Frank Arute, Kunal Arya et al., "Quantum Supremacy Using Programmable Superconducting Processor," *Nature*, October 23, 2019, www.nature.com/articles/s41586-019-1666-5 (accessed December 10, 2019).

39 Jeffrey Ding, *Deciphering China's AI Dream* (Future of Humanity Institute, University of Oxford, March 2018), www.fhi.ox.ac.uk/wp-content/uploads/Deciphering_Chinas_AI-Dream.pdf (accessed July 10, 2019).

40 Chinese venture capital investment in US AI companies between 2010 and 2017 totaled circa US$1.3 billion. China participated in more than 10 percent of all venture deals in 2015, focusing on critical dual-use technologies such as AI, robotics, autonomous vehicles, financial technology, virtual reality, and gene editing. Michael Brown and Pavneet Singh, "How Chinese Investments in Emerging Technology Enable a Strategic Competitor to Access the Crown Jewels of US Innovation," *DIUX*, January 2017 https://admin.govexec.com/media/diux_chinatechnologytransferstudy_jan_2018_(1).pdf (accessed July 10, 2019).

41 Paul Mozur and Jane Perlez, "China Bets on Sensitive US Start-Ups, Worrying the Pentagon," *New York Times*, March 22, 2017, www.nytimes.com/2017/03/22/technology/china-defense-start-ups.html (accessed July 10, 2019).

42 US concerns relate a widespread belief that many Chinese companies can readily be recruited (or 'forced') by Beijing to transfer technology to the Chinese military to bolster its capabilities; they are adjuncts of the Chinese state. Kinling Lo, "China Says the US Claims it Uses Forced Technology Transfer to Boost Military are 'Absurd,'" *South China Morning Post*, January 16, 2019,

www.scmp.com/news/china/military/article/2182402/china-says-us-claims-it-uses-forced-technology-transfer-boost (accessed July 10, 2019).
43 Office of the Secretary of Defense, *Annual Report to Congress: Military and Security Developments Involving the People's Republic of China, 2019* (Washington, DC: US Department of Defense, 2019), https://media.defense.gov/2019/May/02/2002127082/-1/-1/1/2019_CHINA_MILITARY_POWER_REPORT.pdf, p. 48 (accessed July 10, 2019).
44 In 2018, the US lobbied its allies to ban Huawei from building its next generation of mobile phone networks. In response, the UK, Germany, Australia, New Zealand, and Canada have either placed prohibitions on dealing with Huawei on national security grounds or conducted reviews on whether to do so.
45 Lui Zhen, "Why 5G, a Battleground for the US and China, is Also a Fight for Military Supremacy," *South China Morning Post*, January 1, 2019, www.scmp.com/news/china/military/article/2184493/why-5g-battleground-us-and-china-also-fight-military-supremacy (accessed July 10, 2019).
46 Johnson, "The End of Military-Techno *Pax Americana*?"
47 "Xi Jinping's Report at the 19th Chinese Communist Party National Congress," *Xinhua*, October 27, 2017, www.xinhuanet.com//politics/19cpcnc/2017-10/27/c_1121867529.htm (accessed July 10, 2019).
48 Chinese debate on AI and future warfare has increased considerably in recent years. Prominant Chinese research institutions include the PLA Academy of Military Science, National Defense University, and the National University of Defense Technology. For example, see "Liu Guozhi, Artificial Intelligence Will Accelerate the Process of Military Transformation," *Xinhua*, March 8, 2017, www.xinhuanet.com//mil/2017-03/08/c_129504550.htm (accessed July 10, 2019).
49 "Opinions on Strengthening the Construction of a New Type of Think Tank with Chinese Characteristics," *Xinhua*, January 21, 2015, www.chinadaily.com.cn/china/2014-10/27/content_18810882.htmwww.xinhuanet.com/english/download/Xi_Jinping%27s_report_at_19th_CPC_National_Congress.pdf (accessed July 10, 2019).
50 China's most recent five-year plan reportedly committed over US$100 billion to AI. Moreover, as China moves forward with its One Belt One Road-related projects that extend to potentially more than eighty countries, AI would become an integral part of these international infrastructure projects. Wen Yuan, "China's 'Digital Silk Road:' Pitfalls Among High Hopes," *The Diplomat*, November 3, 2017, https://thediplomat.com/2017/11/chinas-digital-silk-road-pitfalls-among-high-hopes/ (accessed July 10, 2019).
51 For example, in quantum computing, China has made significant efforts to integrate its quantum computing and AI research to boost computer AI power and achieve 'quantum supremacy' or the point at which a quantum computer is capable of outperforming a traditional computer. Chinese researchers have claimed to be on track to achieve 'quantum supremacy' as soon as 2019. See Elsa B. Kania and John K. Costello, *Quantum Hegemony? China's Ambitions and the Challenge to US Innovation Leadership* (Washington, DC: CNAS, July 2015), p. 4.

52 Chinese state-directed investment channels include programs such as the Laboratory of National Defence Technology for Underwater Vehicles, the Project for National Key Laboratory of Underwater Information Processing and Control, the National Key Basic Research and Development Programme, the China Aviation Science Foundation, the National Science and Technology Major Project, the National 973 Project, the National Key Laboratory Fund, the National 863 High-Tech Research and Development Programme, and the Ministry of Communications Applied Basic Research Project. Lora Saalman, "China's integration of neural networks into hypersonic glide vehicles," in Nicholas D. Wright (ed.), *AI, China, Russia, and the Global Order: Technological, Political, Global, and Creative Perspectives*, Strategic Multilayer Assessment Periodic Publication (Washington, DC: Department of Defense, December 2018).
53 For a summary of China's tech transfer activities, see Sean O'Connor, "How Chinese Companies Facilitate Technology Transfer from the United States," US–China Economic and Security Review Commission, May 2019, www.uscc.gov/sites/default/files/Research/How%20Chinese%20Companies%20Facilitate%20Tech%20Transfer%20from%20the%20US.pdf (accessed July 10, 2019).
54 Jessi Hempel, "Inside Baidu's Bid to Lead the AI Revolution," *Wired*, December 6, 2017, www.wired.com/story/inside-baidu-artificial-intelligence/ (accessed February 20, 2021).
55 Bill Gertz, "China Reveals Plans for 'Phantom' Underwater Drone War Against the US," *Freebeacon*, November 2, 2018, www.realcleardefense.com/2018/11/06/china_reveals_plans_for_lsquophantomrsquo_underwater_drone_war_against_us_305082.html (accessed December 10, 2019).
56 "Drone Swarming Technique May Change Combat Strategies: Expert," *Global Times*, February 14, 2017, www.globaltimes.cn/content/1032741.shtml (accessed December 10, 2019).
57 G.S. Li, "The Strategic Support Force is a Key to Winning Throughout the Course of Operations," People's Daily Online, January 5, 2016, http://military.people.com.cn/n1/2016/0105/c1011-28011251.html (accessed December 12, 2018).
58 For example, see An Weiping, "Quantum Communications Sparks Off Transformation in the Military Domain," *PLA Daily*, September 27, 2016, https://jz.chinamil.com.cn/n2014/tp/content_7278464.htm (accessed December 10, 2019).
59 Besides, quantum technology might also enable stealth-defeating radar, subatomic lithography for expanded data storage, and advanced scientific modeling and simulation. See Jon R. Lindsay, "Demystifying the Quantum Threat: Infrastructure, Institutions, and Intelligence Advantage," *Security Studies* (2020), www.tandfonline.com/doi/abs/10.1080/09636412.2020.1722853 (accessed February 20, 2021).
60 Kania and Costello, *Quantum Hegemony?*, p. 18.
61 For example, in April 2018, the Chinese People's Political Consultative

Conference convened a symposium to support government engagement with AI leaders from academia and industry. The debate focused on the gap between the degree of AI development in China compared to the US and was supportive of a market-orientated strategy with strong state backing, Francois Godement, *The China Dream Goes Digital: Technology in the Age of Xi* (Paris: European Council on Foreign Affairs, 2018) pp. 1–5.

62 For example, when Google acquired DeepMind, it explicitly prohibited the use of its research for military purposes. Loren DeJonge Schulman, Alexandra Sander, and Madeline Christian, "The Rocky Relationship Between Washington & Silicon Valley: Clearing the Path to Improved Collaboration" (Washington, DC: CNAS, July 2015).

63 Jeremy White, "Google Pledges Not to Work on Weapons after Project Maven Backlash," *The Independent*, June 7, 2018, www.independent.co.uk/life-style/gadgets-and-tech/news/google-ai-weapons-military-project-maven-sundar-pichai-blog-post-a8388731.html (accessed December 10, 2019).

64 Adm. Harry B. Harris Jr. et al., "The Integrated Joint Force: A Lethal Solution for Ensuring Military Preeminence," *Strategy Bridge*, March 2, 2018, https://thestrategybridge.org/the-bridge/2018/3/2/the-integrated-joint-force-a-lethal-solution-for-ensuring-military-preeminence (accessed December 10, 2019).

65 "Advancing Quantum Information Science: National Challenges and Opportunities," *National Science and Technology Council* (Washington, DC: US Department of Defense, July 22, 2017), https://Quantum_Info_Sci_Report_2016_07_22 final.pdf (accessed December 10, 2019).

66 US Department of Defence, "Summary of the 2018 Department of Defense Artificial Intelligence Strategy," February 2019, https://media.defense.gov/2019/Feb/12/2002088963/-1/-1/1/SUMMARY-OF-DOD-AI-STRATEGY.PDF (accessed September 10, 2019).

67 For more on this idea see Andrew Kennedy and Darren Lim, "The Innovation Imperative: Technology and US–China Rivalry in the Twenty-First Century," *International Affairs*, 94, 3 (2017), pp. 553–572.

68 A distinction exists between the erosion of US advantages in ancillary applications based on dual-use AI technologies, and in military-specific AI applications. Where the US retains an unassailable edge in military capacity and innovation, the actual 'threat' posed to the US in the military-technological sphere is less immediate than in general-use AI. This implies that the 'threat' narrative is centered on perceptions of Beijing's future intentions as its military-use AI matures.

69 Currently, the US retains the upper hand in satellite imagery for data-collection and surveillance. The US has circa 373 earth observation satellites compared to China's 134 – including dual-use and multi-country-operated satellites.

70 William G. Beasley, *Japanese Imperialism 1894–1945* (Oxford: Clarendon Press, 1991).

71 Yujia He, *How China is Preparing for an AI-Powered Future* (Washington, DC: The Wilson Center, June 20, 2017), www.wilsoncenter.org/publication/how-china-preparing-for-ai-powered-future (accessed December 12, 2018).

72 Dominic Barton and Jonathan Woetzel, *Artificial Intelligence: Implications for China* (New York: McKinsey Global Institute, April 2017).
73 John R. Allen and Amir Husain, "The Next Space Race is Artificial Intelligence," *Foreign Policy*, November 3, 2017, https://foreignpolicy.com/2017/11/03/the-next-space-race-is-artificial-intelligence-and-america-is-losing-to-china/ (accessed December 10, 2019).
74 Ibid.
75 For example, South Korea has developed a semi-autonomous weapon system to protect the demilitarized zone from North Korean aggression (the SGR-A1), Singapore's 'AI Singapore' is a commercially driven US$110m effort to support AI R&D, and France and the UK have each announced major public-private AI-related initiatives.
76 IR power transition theory recognizes that economic resources are a critical foundation of military power, especially in the context of a declining hegemon. See A.F.K. Organski and Jacek Kugler, *The War Ledger* (Chicago and London: University of Chicago Press, 1980); and Robert Gilpin, *War and Change in World Politics* (Cambridge: Cambridge University Press, 1981).
77 It has been argued that the more stratified hierarchy that exists under bipolarity, where two poles have a clear preponderance of power, middle and small states are less likely to experience status dissonance and dissatisfaction associated with a multipolar order. William Wohlforth, "US strategy in a unipolar world," in John G. Ikenberry (ed.), *America Unrivaled: The Future of the Balance of Power* (Ithaca, NY: Cornell University Press, 2002), pp. 98–121.
78 The historical record has demonstrated that great military powers often struggle to adapt and assimilate new technologies into military organizations, and disruptive technological adaptations frequently result in significant organizational and creativity-stifling pressures. Michael C. Horowitz, *The Diffusion of Military Power: Causes and Consequences for International Politics* (Princeton, NJ: Princeton University Press, 2010).
79 Gregory D. Koblentz, *Council Special Report-Strategic Stability in the Second Nuclear Age* (New York: Council on Foreign Relations Press, 2014).
80 See Peter W. Singer, *Wired for War: The Robotics Revolution and Conflict in the 21st Century* (New York: Penguin, 2009).
81 In addition to mimicry and espionage, several other factors influence the diffusion of military technologies, including system complexity, unit costs, and whether the underlying technology comes from commercial or military-use research. See Andrea Gilli and Mauro Gilli, "Why China Has Not Caught Up Yet," *International Security* 43, 3 (2019), pp. 141–189; and Horowitz, *The Diffusion of Military Power*.
82 The hardware costs and computer power associated with sophisticated software required to support 'narrow AI' applications are potentially significant – the more complex the algorithm required to train AI applications; the more computational power is needed. Robert D. Hof, "Deep Learning," *MIT Technology Review*, April 23, 2013, www.technologyreview.com/s/600989/man-and-machine/ (accessed December 10, 2019).

83 Several experts have argued that developing technically advanced weapon systems has become dramatically more demanding, making internal balancing against the US more difficult. Stephen G. Brooks, *Producing Security: Multinational Corporations, Globalization, and the Changing Calculus of Conflict* (Princeton, NJ: Princeton University Press, 2006); Jonathan D. Caverley, "United States Hegemony and the New Economics of Defense," *Security Studies*, 16, 4 (October–December 2007), pp. 598–614; and Andrea Gilli and Mauro Gilli, "The Diffusion of Drone Warfare? Industrial, Organizational and Infrastructural Constraints," *Security Studies* 25, 1 (2016), pp. 50–84, and Gilli and Gilli, "Why China Has Not Caught Up Yet," pp. 141–189.

84 For example, the shift in British military fortunes in the early twentieth century was a product of not only the transformation of military technology, but also the coalescence of British society and military culture, and the demands of industrial-age warfare on the other. For the intersection of strategic culture, military doctrine, and broader societal factors see Iain Alastair Johnstone, "Thinking About Strategy Culture," *International Security*, 19 4 (1995), pp. 32–64; and Elizabeth Kier, "Culture and Military Doctrine: France between the Wars," *International Security*, 19, 4 (1995), pp. 65–93.

85 Stephen Biddle, *Military Power: Explaining Victory and Defeat in Modern Battle* (Princeton, NJ: Princeton University Press, 2006).

86 James E. Tomayko, *Computers Take Flight: A History of NASA's Pioneering Digital Fly-By-Wire Project* (Washington, DC: National Aeronautics and Space Administration, 2000), pp. 24–25, and 30.

87 Kareem Ayoub and Kenneth Payne, "Strategy in the Age of Artificial Intelligence," *Journal of Strategic Studies* 39, 5–6 (2016), pp. 793–819, p. 809.

88 US Department of Defense, "Task Force Report: The Role of Autonomy in DoD Systems" (Washington, DC: US Department of Defense, July 2012), pp. 46–49.

89 E. Gray et al., "Small Big Data: Using Multiple Data-Sets to Explore Unfolding Social and Economic Change," *Big Data & Society* 2, 1 (2015), pp. 1–6.

90 As of November 2019, seven of the most powerful computers are in the US, and three in China. However, 219 of the top 500 supercomputers are in China – 116 are in the US *Top 500 The List*, November 2018. TOP500 (Lists, top500 list (excel), www.top500.org/lists/top500/ (accessed December 10, 2019).

91 This notion of bipolar competition could be accentuated in the case of 'general' AI (or 'superintelligence'), as opposed to 'narrow' use AI. In theory, a leader in general AI technology might lock in the advantages gained by the development of this technology so large that others are unable to catch up. See Nick Bostrom, *Superintelligence: Paths, Dangers, Strategies* (Oxford: Oxford University Press, 2014).

92 For example, recent reports suggest that much of the data collected by the US military – especially from commercial sources – is considered poorly integrated, siloed, and unusable in many cases. Sydney J. Freedberg Jr., "EXCLUSIVE Pentagon's AI Problem is 'Dirty' Data: Lt. Gen. Shanahan," Breaking Defense, November 13, 2019, https://breakingdefense.com/2019/11/exclusive-penta gons-ai-problem-is-dirty-data-ltgen-shanahan/ (accessed March 10, 2020).

93 Alan Beverchen, "Clausewitz and the non-linear nature of war: systems of organized complexity," in Hew Strachan and Andreas Herberg-Rothe (eds), *Clausewitz in the Twenty-First Century* (Oxford: Oxford University Press, 2007), pp. 45–56.
94 A recent report claimed that the DoD was actively engaged in over 600 AI-related projects. To date, however, analysts assert that the DoD has struggled to shift bottom-up projects into established programs of record, with requisite critical mass for organizational changes. National Security Commission on Artificial Intelligence Interim Report to Congress, November 2019, www.nscai.gov/reports (accessed August 10, 2019).
95 For Chinese views see Yang Feilong and Li Shijiang, "Cognitive Warfare: Dominating the Era of Intelligence," *PLA Daily*, March 19, 2020, www.81.cn/theory/2020-03/19/content_9772502.htm (accessed August 10, 2019).
96 This idea resonates with neorealist IR scholars who posit that the anarchy of the international system forces all great powers to exploit technological change or suffer the risk of not surviving. Thus, the nature of the international system requires states to adapt to new technologies (such as AI) or face threats to their survival as they lag behind other states in relative capabilities. Kenneth Waltz, *Theory of International Politics* (Reading, MA: Addison-Wesley, 1979). For recent Chinese views that echo these sentiments see Yang Feilong and Li Shijiang "Cognitive Warfare."
97 US Department of Defense, *Nuclear Posture Review* (Washington, DC: US Department of Defense, February 2018), p. 14.

# 4

# US–China crisis stability under the AI nuclear shadow

Will AI-augmented technology increase the risk of military escalation between great military rivals?[1] This chapter argues that the fusion of AI-enhanced technologies with both nuclear and conventional capabilities will be destabilizing, and that this problem will be exacerbated by the fact that China and the US have divergent views of the escalation risks of co-mingled (or 'entangled') nuclear and conventional capabilities.[2]

From what we know about the advances in military AI today, AI-augmented military technologies will exacerbate the risks of 'inadvertent escalation,'[3] caused by the co-mingling of nuclear and non-nuclear capabilities, thus undermining strategic stability. As former US Deputy Assistant Secretary of the Bureau of Arms Control, Verification, and Compliance, Frank Rose asserted, "strategic stability encompasses much more than just nuclear relations."[4] As described in chapter 2, *nuclear deterrence* is no longer synonymous with *strategic* deterrence or stability.[5]

Based on an understanding of how China and the US think about entanglement, escalation, deterrence, and strategic stability, this chapter elucidates the potential risks of unintentional (inadvertent or accidental) escalation and conflict caused by these dynamics. In this case, escalation dynamics and deterrence refer to mechanisms by which armed conflict can begin, how seemingly minor skirmishes can be inflated, and how these processes might be controlled or managed to make the outbreak, escalation, and termination of conflict less likely.[6]

Towards this end, the chapter unpacks the following research puzzles. How might AI and other advanced or emerging technologies exacerbate the co-mingling problem-set? How concerned is China about inadvertent or accidental escalation? How serious are the escalation risks arising from entanglement in a US–China crisis or conflict scenario? How can the US, China, or others mitigate the escalation risks exacerbated by the advent of AI technologies? In this way, the chapter sketches a roadmap for a journey that US and Chinese leaders hope never to embark upon but must nonetheless prepare for.[7]

The remainder of the chapter proceeds in three sections. The first identifies the key drivers and potential new pathways to escalation – via misperception, the security dilemma spiral, increasing an adversary's propensity to take greater risks, and the 'fog of war' – created by the co-mingling of nuclear and advanced strategic non-nuclear technologies capabilities – or conventional capabilities that can compromise nuclear weapons. It then uses this framework to illustrate how the divergent Chinese and US views of entanglement exacerbate each of the pathways to escalation.

The second section describes the divergent US–China strategic approaches to escalation, entanglement, deterrence, and strategic stability. It uses evidence from the US and Chinese strategic views and approaches to illustrate these potential pathways. Then it describes how AI and related technologies exacerbate some or all of these dangers in the case of a US–China conflict.

The final section considers possible ways to mitigate the risks of inadvertent escalation in the context of emerging technology like AI between nuclear-armed great powers. It concludes that the inherently destabilizing characteristics of AI, coupled with the multifaceted interplay of AI with strategic weapons and a competitive geopolitical environment, will make these risks extremely challenging to manage.

## Advanced military technology and escalation pathways

How might AI and other advanced or emerging technologies exacerbate the co-mingling problem? Four trends associated with military technology development have amplified the escalation risks of the co-mingled nuclear and non-nuclear capabilities.[8] First, technological advances in conventional long-range precision missiles, command, control, communications, and intelligence (C3I) and early warning systems, cyber-weapons, autonomous weapon systems, and military AI tools. Second, a growing military reliance on global information infrastructure supplied by dual-use C3I capabilities – especially in space and cyberspace. Recent reports suggest that China (and Russia) believes the US intends to leverage AI to enhance its conventional counterforce capabilities – especially AI-enhanced intelligence systems – which might undermine the survivability of its nuclear forces.[9] Third, and paradoxically, the increasing vulnerability caused by enhancements to these critical dual-use enabling systems is especially so from cyber-attacks (see chapter 7). Finally, the emergence of conventional warfighting military doctrines – China, Russia, and the US – emphasizes surprise, speed, deep strikes, and pre-emption to bolster deterrence and achieve 'escalation dominance.'[10]

Taken together, these trends could unintentionally (i.e. inadvertently or accidentally) escalate a conventional war to a nuclear confrontation in several ways. First, the multifunctional intended targets of dual-use (with civilian and military uses) technologies and dual-payload capabilities – armed with nuclear or conventional warheads – might cause misperceptions and miscalculations between adversaries triggering inadvertent escalation.[11] Deterrence theorists posit that conventional counterforce capabilities can undermine the credibility of states' nuclear deterrence forces, creating strike first incentives.[12] Further, an adversary may misperceive a conventional strike as a conventional counterforce campaign, or as a prelude to a nuclear first strike, and, in turn, inadvertently escalate a situation.[13]

Second, divergent views held by adversaries about the intended use, potential impact, and the circumstances under which particular military technologies (especially dual-use) may be employed, could trigger a negative action–reaction dynamic known as a security dilemma.[14] Human cognition, and thus effective deterrence, is predicated on reliable and unambiguous information. Consequently, if an adversary is concerned that the information available to them is limited (or worse inaccurate), they will likely assume the worst and act accordingly.[15] Because of this condition in international relations, great powers are inclined to strengthen their deterrence and signal resolve to counter the perceived expansionary or aggressive (i.e. not solely motivated by legitimate security interests) intentions of strategic rivals.[16]

Third, the development and use of certain types of strategic non-nuclear weapons and enabling technologies (e.g. hypersonic weapons, cyber, missile defense, quantum computing, autonomous weapons, and AI-augmented systems), might influence an adversary's attitude to risk during a crisis, thereby making it more (or less) prone to escalate a situation. As demonstrated in chapter 3, technologies like AI, cyber, hypersonic vehicles, and autonomous weapons can create (or amplify) the perception of first-mover advantages, which can increase inadvertent escalation risks.[17] Today, there is a genuine risk among great military powers (Russia, China, and the US) that non-nuclear attacks on an adversary nuclear deterrent and their attendant command and control systems – especially from cyber-attacks – could escalate a situation to nuclear warfare.[18] For instance, false-flag operations (i.e. from state proxies or third party actors) in cyberspace can sow fear and uncertainty in actors' minds that their ability to retaliate is weakened (see chapter 8) – above all, the fear of losing the first-mover advantages afforded by these capacities.

Finally, non-nuclear enabling technologies such as AI machine learning (ML) and cyber capabilities employed during a crisis might affect the uncertainties caused by situational awareness during a conflict – or the 'fog of war' – that could increase or decrease escalation risks.[19] The heightened sense of vulnerability caused by the uncertainties of warfare could make a state *more* risk-averse, and thus, more prone to escalate a situation.[20] For example, a cyber-attack that succeeds in compromising an adversary's NC3 systems could cause either or both sides to overstate (or understate) its retaliatory capabilities, and thus be more inclined to escalate a situation (chapter 8 examines the AI-cyber security threat nexus in detail).[21]

Restricted information flows between adversaries – and within military organizations – might cause 'asymmetric information,' thereby thickening the fog of war.[22] Rather than reducing the uncertainties associated with warfare, therefore, AI-augmentation might reduce the quality and flow of information during wartime. Further, suppose AI and autonomy allow (or are perceived to give) either side in a competitive dyad to achieve an asymmetric military – conventional or nuclear – advantage. In that case, the weaker side may feel compelled to rely more heavily on its nuclear capabilities for deterrence.[23] It may even adopt nuclear warfighting.[24] Thus, as states continue to augment their forces with AI, the melding of new and legacy systems operating in tandem could make it more challenging to understand human–machine interactions.[25] If states incorrectly predict the outcome of this mix in less observable fields such as AI, therefore, these capabilities could be exaggerated or understated.[26] How an adversary intends to use AI-enabled capabilities – and under what circumstances – will likely remain unknown, even if the algorithms and data sources used are known to the other side.[27]

Research on bargaining and war also suggests that uncertainty about states' capabilities and intentions caused (or exacerbated) by opacity makes it more difficult to reach diplomatic solutions once a crisis begins.[28] Asymmetric information between rivals about military capabilities – and uncertainty about their intentions – means one or both sides might assume they will triumph if a situation spirals.[29] Because the availability (and reliability) of information about clandestine capabilities will likely be limited,[30] these capabilities can influence the military balance and lead to crisis bargaining failure and war.[31] Further, opacity about the military balance caused by clandestine capabilities may increase an actor's risk tolerance, thereby creating overconfidence (or false optimism) in its ability to win a conflict and prevent loss.[32]

During peacetime competition, states' willingness to reveal their clandestine capabilities will likely be determined by the uniqueness of specific capabilities, and the anticipation that an adversary will be able to develop countermeasures successfully. Where a capability is perceived as less unique, and where countervails to these capabilities by adversaries is considered less likely, states are more likely to keep their capabilities secret.[33] Because of the opacity surrounding AI systems – especially the black box issue associated with ML algorithms described in chapter 1 – this level of uncertainty may be exacerbated, thus increasing escalation risks during a crisis.[34] In sum, asymmetric information situations between adversaries about the balance of military power can erode crisis stability and create rational incentives to escalate to nuclear confrontation.

Similar to strategic stability at a broader level, therefore, AI's impact on escalation will also be shaped by broader and more nuanced factors, which influence the trajectory of emerging technologies. These include military strategy, operational concepts, and doctrine (which seeks to manipulate escalation risks), military culture and organizational constructs, alliance structures, and domestic politics and public opinion, to name a few.[35] As shown in chapter 2, the Cold War experience demonstrated that emerging technologies act primarily to enable independent variables, which, combined with broader endogenous factors, can heighten escalation risks between rivals.[36] Therefore, in isolation, technology is not the only (or even necessarily the leading) cause of military escalation.[37] Understanding these broader factors is even more crucial when states hold divergent views on how military technologies like AI should be employed to influence deterrence and escalation.

## Divergent US–China thinking on military escalation risk

How could divergent Chinese and US views of entanglement exacerbate the pathways to escalation caused by AI and emerging technology? While US defense analysts and their Chinese counterparts are aware of the potential escalation risks between nuclear-armed great powers, their respective doctrines do not address how an adversary might respond to (i.e. deter or suppress) escalatory behavior. Instead, these rival strategic communities generally assume that escalation in future conflict can be effectively countered and contained by establishing and sustaining escalation dominance.[38] Like the US, the Chinese doctrinal emphasizes seizing the initiative early and pre-emptively in conventional confrontation to achieve escalation dominance.[39] This approach might, however, risk triggering rapid and possibly uncontrollable escalation to a nuclear level.[40] Remove commonly held escalation thresholds and a mutual framework to deter either side violating of them, and a US–China crisis operating under the assumption that *both* sides can effectively control escalation – for example, in the South China Seas, the Korean Peninsula, or Taiwan Straits – would likely increase the risks of inadvertent escalation.[41]

It is noteworthy that China and the US share a similar holistic approach to the interoperability of emerging technologies – including AI – in conventional strikes. This approach is reflected in their respective operating concepts and force structures.[42] This interoperability, together with states' increasingly co-mingled nuclear and non-nuclear capabilities, creates a *de facto* escalation framework for nuclear deterrence.[43] That is, the risk of escalation to more serious retaliation increases the potential costs through low-level attacks below the nuclear threshold, such as conventional counterforce attacks.[44]

The sequence of events triggering inadvertent escalation might unfold as follows: state A attacks state B's early warning systems located in B's non-nuclear launch sites as a tactical maneuver. Because these systems are also used by B to warn them of imminent nuclear attacks, B concludes the attacks are a prelude to a disabling first strike.[45] Thus, action A intended to remain under the nuclear threshold, B views as deliberate escalation. That is, B assumes A *must know* that an attack on its dual-use systems would be highly escalatory. Because A does not consider its opening salvo as escalatory, it views B's response as highly provocative, triggering unintentional escalation. For example, during the Cold War, increasing US counterforce capabilities, coupled with US and Soviet nuclear doctrines that went beyond the assured destruction mission, increased the likelihood of pre-emptive attacks by both sides on the other's command and control capabilities. In a modern context, this approach is also at the core of the often criticized US Department of Defense's AirSea Battle Concept.[46]

Not all escalation is unintentional or accidental, however.[47] The inadvertent risks illustrated in this fictional scenario could equally be applied to states' use of technology to pursue intentional escalation. For instance, an actor might deliberately manipulate or subvert an adversary's NC3 networks and erode their nuclear deterrence capacity as a means of coercion during a crisis on conflict, to force their adversaries to capitulate more rapidly than would otherwise be the case, or even to persuade a rival not to resort to military force in the first place.[48] With the reduced direct risk to military personal, AI-enabled autonomous systems might increase the willingness of adversaries to employ these options and escalate a situation.[49] Further, because of an entrenched belief held by the Chinese, Russian, and US strategic communities that a military escalation is a *deliberate act* (i.e. not unintentional or inadvertent), escalatory risk might be underestimated, and thus, amplified.[50]

An additional complicating factor is that adversaries are understandably reluctant to broadcast their military vulnerabilities. As a result, in the lead-up to a crisis or conflict, inadvertent escalation risks may go undetected.[51] Moreover, if a state perceives itself as vulnerable (i.e. against a superior military power), they may develop and deploy capabilities designed to increase the chances of *accidental* escalation deliberately to amplify the deterrent effect of a particular capability.[52] For example, the credible threat posed by the deployment of untested and unverified AI-enabled drone swarms (see chapter 6) into a complex combat zone might convince a target to eschew belligerence for prudence and exhibit tactical patience – analogous to Schelling's "threat that leaves something to chance."[53] Conversely, when the lives of military personnel are not directly at stake, commanders may be less concerned about breaching escalation thresholds and more inclined to accept greater risk as a crisis unfolds (chapter 6 will examine the strategic impact of AI-enhanced drone swarms).

Escalation thresholds that represent the various rungs of intensity (or the 'escalation ladder'[54]) in a confrontation are socially constructed, and thus, inherently subjective.[55] Consequently, the inability to anticipate the escalation effects of a particular action may be due to misperceptions caused by the failure to recognize the second- or third-order consequences of a particular situation or action, or both.[56] For example, what happens when some (or all) of intelligence used for signaling and decision-making – on both sides – is no longer human? Chapter 8 will return to this issue. While it remains axiomatic that human decisions escalate (or de-escalate) a situation – not weapon systems per se – military technology that enables offensive capabilities to operate at higher speed, range, and lethality will likely move a situation more quickly up the escalation rungs to a strategic level of conflict.[57]

New escalation thresholds and operating norms for AI-augmented weapons have yet to emerge. Today's thresholds in the context of autonomous weapon systems are considered inappropriate and ambiguous.[58] Without commonly held operational norms and an adversary's strategic priorities and political objectives, militaries deploying military AI could inadvertently cross escalation thresholds.[59] In 2016, China captured a US unmanned undersea vehicle, asserting it posed a hazard to Chinese maritime navigation. In response, Washington called China's behavior 'unlawful', claiming the vehicle was a 'sovereign immune vessel.'[60] This diplomatic episode demonstrated the potential risk of inadvertent escalation caused by the ambiguity surrounding the deployment of new (and especially dual-use) technology in contested territory between strategic rivals. Recent history demonstrates how these escalatory dynamics could play out with unmanned systems (more on this idea in chapter 7).[61] In sum, the combination of first-strike vulnerability and opportunity enabled by a growing portfolio of technologies for warfighting, coercion, and influence, such as AI, will have significant implications for escalatory dynamics in future warfare.[62]

## Chinese sanguine attitude to escalation

Managing military – especially inadvertent – escalation risk has not been a traditional feature of Chinese strategic thinking.[63] China's strategic community is believed to share a high level of confidence in the ability of China's long-standing no-first-use (NFU) nuclear pledge to control escalation.[64] Because Chinese analysts view China's NFU commitment as a *de facto* firebreak between its conventional and nuclear capabilities to de-escalate a situation, the resultant overconfidence might increase inadvertent escalation risks. The relative optimism of Chinese analysts can also be attributed to the belief that any escalation to nuclear threats or use would be *intentional and belligerent*.[65] This overconfidence could make it less likely that Chinese leaders recognize escalation risks caused by miscalculation or misperception of US intentions. China's sanguine attitude to managing escalation can, in part, be attributed to the belief that either side cannot easily control nuclear weapons once nuclear Rubicon is crossed. Because of this belief, Chinese analysts do not believe that a limited nuclear war would stay limited. Moreover, China's operational doctrine does not contain plans to wage a limited nuclear war, which China might pursue if it believed nuclear escalation could be controlled.[66]

Faced with the prospect of an imminent US conventional counterforce attack, a less robust second-strike capacity could heighten Beijing's sense of vulnerability, thereby limiting any potentially de-escalatory effects China's nuclear pledge might have otherwise had.[67] Put another way, escalation may happen inadvertently if a shift occurs in the risk–reward trade-off associated with the intense coercive bargaining with conventional weapons – to signal resolve or to pre-empt an attack below the nuclear threshold, which is central to bargaining theory.[68] For example, Chinese analysts view hypersonic conventional cruise missiles (see chapter 6) as an effective means to penetrate US missile defenses, thereby strengthening China's nuclear deterrence.[69]

According to Professor of Political Science at the University of Pennsylvania Avery Goldstein, Chinese overconfidence in its ability to prevent conventional military confrontation escalating to nuclear war might increase the risk of a conflict or crisis inadvertently or accidentally crossing the nuclear Rubicon.[70] Moreover, divergent US–China attitudes about controlling escalation – both above and below the nuclear threshold – could also be detrimental to crisis stability. The US defense community is generally more concerned than its Chinese counterparts that a low-level conventional conflict could escalate to a strategic level, but less worried about the ability of US forces to control escalation *above* the nuclear threshold.[71] Paradoxically, therefore, during a US–China crisis or conflict, Washington may *overstate* the possibility that Beijing would use nuclear weapons, and simultaneously, *understate* the scale of a Chinese retaliatory nuclear response. In short, divergent US–China attitudes about de-escalating a low-intensity conventional make a conventional conflict more likely to escalate to high intensity.[72]

Fearful that a US conventional counterforce attack might circumvent its deception and dispersal tactics, and upend the survivability of its nuclear arsenals, Chinese leaders would be under intense pressure to engage in limited nuclear escalation to reclaim escalation dominance.[73] Perhaps Beijing's confidence might embolden the US to gamble that a limited nuclear first strike against China would not provoke a nuclear response, compelling Chinese leaders retaliate to protect its second-strike capability and increasing nuclear escalation risks.[74] Further, US–China crisis management could be complicated because Washington and Beijing do not have the protracted and eventful crisis management history experienced by the Soviets and the US during the Cold War-era.

Furthermore, in a US–China conflict, the US would have a strong incentive to pre-emptively attack China's mobile missiles and attendant (and likely dual-use) C3I systems to achieve escalation dominance, which Beijing could (mis)perceive as a conventional counterforce attack, or worse, as a precursor to a first nuclear strike.[75] Chinese analysts generally assume that the US intends to undermine China's relatively small nuclear deterrent and attendant support systems with advanced conventional weapons – especially US conventional prompt global strike and missile defenses.[76]

Chinese military writings remain under-theorized on a range of critical issues about Chinese thinking (or even awareness) about the inadvertent escalation risks associated with the entanglement. Chinese analysts recognize that a significant threshold exists between nuclear and conventional warfare, but Chinese military writings do not explicitly address the mechanisms for moving through the various rungs of a notional escalation.[77] Chinese open-source literature suggests that escalation redlines would likely include the warning of an imminent nuclear or conventional attack on China's nuclear forces, and its co-mingled (and likely co-located) C3I systems.[78] Chinese strategists do not appear concerned that this ambiguity might be mistaken for China's preparation to use its nuclear weapons.[79]

Furthermore, while it is unlikely that China has deliberately entangled its nuclear and conventional capabilities to reduce the vulnerability of its non-nuclear forces against a US strike, the strategic ambiguity from co-mingling its warheads by protecting China's nuclear arsenals – against a US pre-emptive attack – could be advantageous nonetheless.[80] Moreover, during a crisis, Beijing's efforts to obfuscate its missile operations (i.e. disperse its nuclear and conventional forces simultaneously) would reduce the

US ability to track Chinese missiles after leaving their garrisons – a concern raised by the US Department of Defense.[81] However, this strategic ambiguity might persuade Beijing against the strategic rationale of separating its forces to lower the risk of inadvertent escalation.[82] A critical challenge for external analysts is to identify the escalation thresholds – both Chinese thresholds and China's perception of US thresholds – that, once crossed, would be viewed by Beijing as justification for a nuclear response without violating NFU.

Experts on Chinese nuclear forces maintain that as long as China continues its rudimentary deception and dispersal tactics, it is highly improbable the US would be able to develop the capacity to locate all of China's land-based mobile missiles.[83] Moreover, Chinese analysts believe that a combination of existing technical limitations of US intelligence, surveillance, and reconnaissance (ISR) systems, and the basing modes (i.e. hardened underground facilities) housing these strategic assets, would prevent the US from successfully executing a disarming first strike with conventional (or nuclear) weapons.[84] Were advances in the accuracy and remote sensing of US counterforce capabilities enabled by AI systems to undermine this confidence, however, the implications for the survivability of Chinese mobile missiles and, in turn crisis stability, could be severe.[85]

Open sources indicate that Chinese analysts tend to overlook the impact of the 'fog of war' on decision-making and escalation. For example, Chinese experts have opined that in future, warfare through processing and analyzing intelligence at machine-speed, AI-powered ML applications might disperse the 'battlefield fog.' For example, Chinese analysts believe that through AI ML techniques (the so-called 'algorithm game') reducing the 'fog of war' and enhancing situational awareness, commanders will increase the speed and accuracy of predicting battlefield scenarios and, in turn, 'winning before the war.'[86] As a corollary, Chinese analysts generally hold a relatively sanguine view on AI-enabling advances in ISR systems, to mitigate escalation risks during a crisis.[87] While intelligence gathering that sheds light on an adversary's intentions and capabilities can provide reassurances that may improve stability, intelligence collection (e.g. cyber espionage targeting an adversary's NC3 systems), particularly during a crisis, can also produce misperceptions and miscalculations that have escalatory effects (chapters 8 and 9 develop this theme further).

## Mitigating and managing escalation risk

What can the US, Chinese, or others do to mitigate the escalation risks exacerbated by the advent of AI technologies? Efforts to mitigate the inadvertent escalation risk depends on analysis and understanding of an adversary's capabilities and intentions, as well as empathy – how others might view one's actions and particular situations as escalatory.[88] Several conditions are necessary to successfully anticipate and manage inadvertent escalation risks, including:[89] anticipating possible escalation paths and clarifying the thresholds that might occur in a particular situation;[90] educating defense planners to the nature and probability of inadvertent escalation and the potential risks in particular combat scenarios; effectively communicating to their leaders potential inadvertent escalation risks they may not have recognized or fully comprehend;[91] and understanding how an adversary might perceive and interpret events that have not yet occurred and are subject to change.[92]

Unfortunately, psychologists generally consider humans as poor predictors of risk – in particular, risk assessments made in the context of feedback loops (i.e. action and counter-reaction cycles) and in situations where humans have limited prior knowledge or experience.[93] As shown in chapter 1, AI ML techniques exhibit *both* of these characteristics. Further, given the centrality of political and strategic contexts in which states develop and deploy technology with escalatory effects, any attempts to mitigate and control these risks will need to be cognizant of these broader exogenous drivers. While efforts to control, constrain, or prohibit even technologies like AI may be laudable and warranted, conceived in isolation from a broader political and strategic context, they will not necessarily reduce escalation risks.

Today's competitive multipolar nuclear order will make it unlikely that improvements in intelligence collection, analysis, and situational awareness from advances in AI will have stabilizing effects.[94] With nine nuclear-weapon states and multiple strategic rivalries that directly and indirectly interact with one another within the web of interlocking deterrence dyad – and triads and beyond – the virtuous cycle that might flow from enhanced reassurances and confidence-building measures premised on comprehensive intelligence would require information symmetry (i.e. equal access to intelligence and analysis systems) between rivals, together with shared confidence in the accuracy and credibility of these systems. Perhaps most challenging of all, in a world of revisionist and dissatisfied nuclear-armed states, all parties' intentions would need to be genuinely benign for this optimistic scenario to occur.[95] According to scholar John Mearsheimer, "as long as the system remains anarchic, states will be tempted to use force to alter an unacceptable status quo."[96]

Against this competitive geopolitical backdrop, one state's effort to reduce the risks of inadvertent escalation by signaling a reluctance to risk warfare at a nuclear level of conflict (or signaling intent), may embolden an adversary to engage in deliberate escalation *below* the nuclear threshold – known as the 'stability–instability paradox.'[97] These dynamics will make states more inclined to assume the worst of others' intentions, especially in situations where the legitimacy of the status quo is contested (i.e. maritime Indo-Pacific). In turn, one state's efforts to enhance the survivability of its strategic forces will likely be perceived by an adversary as a threat to its nuclear retaliatory capability – or second-strike capacity.[98] Furthermore, if either or both sides believe that national security and regime survival faces an existential threat, war termination will become more problematic.[99]

## Conclusion

This chapter argued that divergent US–China views on the co-mingling of nuclear and strategic non-nuclear capabilities will exacerbate the destabilizing effects of the synthesis of these capabilities with AI in future warfare. Further, the combination of first-strike *vulnerability* (or strategic force vulnerability) and *opportunity* (or first-mover advantage) enabled by advances in military technologies like AI will have destabilizing implications for escalation dynamics in future warfare.[100]

The co-mingling of nuclear and advanced conventional military technologies could inadvertently escalate a conventional war to a nuclear confrontation in several ways. First, the multifunctional intended targets of dual-use and dual-payload capabilities could cause misperception and misinterpretation between adversaries (i.e. of another's intentions, redlines, and willingness to use force), triggering an inadvertent escalation. Second, the adversaries' divergent views about the intended use, potential impact, and the circumstances under which particular military technologies may be employed could trigger a negative action–reaction escalatory spiral of mistrust, known as a security dilemma. Third, the development and deployment of particular strategic non-nuclear weapons and enabling capabilities, including AI, could influence an adversary's attitude to risk during a crisis, making it more (or less) prone to a *de facto* escalation pathway for nuclear deterrence. Finally, non-nuclear enabling technologies used during a crisis could thicken the fog of war, causing asymmetric information situations between adversaries.

Today, it is axiomatic that humans escalate (or de-escalate) a military situation. However, military technology like AI that enables offensive capabilities to operate at higher speed, range, and lethality, could move a situation more quickly up the escalation rungs, crossing thresholds up to a strategic level of conflict. These escalatory dynamics would be greatly amplified by the development and deployment of AI-augmented tools functioning at machine-speed. As shown in Part I, military AI could push the pace of combat to a point where the actions of machines surpass the cognitive and physical ability of human decision-makers to control or even fully understand, future warfare.

Chinese analysts' apparent high level of confidence in the ability of Beijing's long-standing NFU pledge to avoid escalation might, paradoxically, actually increase the escalation risks caused by the entanglement of its nuclear and non-nuclear assets. Because Chinese analysts view NFU as a *de facto* firebreak between the use of its conventional and nuclear capabilities, this overconfidence could increase inadvertent escalation risks and erode US–China crisis stability. If AI-enhanced ISR enabled US counterforce capabilities to locate China's nuclear assets more precisely, this confidence might be undermined, causing destabilizing use-them-or-lose-them situations.

Inadvertently, escalation might occur with conventional weapons used by Beijing to signal resolve or pre-empt an attack at lower levels of the escalatory ladder; as a result of China's confidence in its ability to control and commitment to dominate – escalation, or its acceptance of the stability–instability paradox. In sum, controlling escalation risks involves human perceptions and behavior, rather than merely a quantitative or engineering problem to achieve a survivable nuclear arsenal.

Chinese military writings remain under-theorized on a range of critical issues about China's thinking on the inadvertent escalation risks associated with nuclear and non-nuclear assets' entanglement. Moreover, there appears to be a lack of understanding (or even recognition) within China's strategic community of the concept of cross-domain deterrence, or how to manage the escalation risks associated with China's co-mingled capabilities. While experts do not believe China has deliberately entangled its nuclear and conventional capabilities to reduce the vulnerability of its non-nuclear forces against a US attack, the strategic ambiguity this coalescence creates could, nonetheless, be advantageous for this precise reason.[101] This ambiguity (deliberate or otherwise) may dissuade Beijing from separating these forces to reduce inadvertent escalation risks, thus blurring conventional and nuclear warfighting.

Given the multifaceted interplay of AI with strategic weapons – both nuclear and conventional weapons with strategic effects – a deep appreciation of the dynamic relationship between AI technology, escalation, and crisis decision-making is needed. Understanding how competing strategic communities view these dynamics, and the implications of these trends for nuclear force structure, nuclear nonproliferation, arms control, escalation management, and cross-domain and extended deterrence, will be increasingly critical task statecraft. Part III uses four case studies to illuminate the strategic implications of fusing AI with a range of advanced military systems.

## Notes

1 Military escalation in this context refers to an increase in the intensity or scope of conflict that crosses thresholds (or ladders) considered significant by one or more parties involved. Forrest E. Morgan et al., *Dangerous Thresholds: Managing Escalation in the 21st Century* (Santa Monica, CA: RAND, 2008), p. xi.
2 'Entanglement' refers to dual-use delivery and support systems and nodes (i.e. early warning systems) that can have both nuclear and non-nuclear capabilities. It can also refer to strategic non-nuclear threats – including AI-augmented weapon systems – to nuclear weapons and their associated command, control, communication, and intelligence systems (examined in chapter 9). See James M. Acton, "Escalation Through Entanglement: How the Vulnerability of Command-and-Control Systems Raises the Risks of an Inadvertent Nuclear War," *International Security*, 43, 1 (2018), pp. 56–99.
3 'Inadvertent escalation' in this context refers to a situation where one state takes an action that it does not believe the other side will (or should) regard as escalatory but occurs unintentionally nonetheless. These risks may exist in peacetime, during a crisis, and in warfare. See Barry R. Posen, *Inadvertent Escalation: Conventional War and Nuclear Risks* (Ithaca, NY: Cornell University Press, 1991).
4 Frank A. Rose, "Ballistic Missile Defense and Strategic Stability in East Asia," February 20, 2015, https://2009-2017.state.gov/t/avc/rls/2015/237746.htm (accessed March 10, 2020).
5 It is noteworthy that the US has historically not placed the goal of strategic stability above sustaining military primacy (or preeminence) and the first-mover advantage vis-à-vis adversaries. During the Cold War, once the Soviets acquired survivable strategic nuclear forces, the US only then settled for stable mutual deterrence – above all, to reduce the risk of inadvertent nuclear war. Aaron R. Miles, "The Dynamics of Strategic Stability and Instability," *Comparative Strategy*, 35, 5 (2016), pp. 423–437.
6 Herbert Lin and Amy Zegart (eds), *Bombs, Bytes and Spies: The Strategic Dimensions of Offensive Cyber Operations* (Washington, DC: Brookings Institution, 2019), p. 11.
7 Beyond the US–China deterrence dyad, other nuclear-armed strategic rivalries would also be subject to similar forms of escalation dynamics described in this chapter, including India–Pakistan, US–Russia, and perhaps US–North Korea.
8 Acton, "Escalation through Entanglement," pp. 56–99.
9 Edward Geist and Andrew Lohn, *How Might Artificial Intelligence Affect the Risk of Nuclear War?* (Santa Monica, CA: RAND Corporation, 2018), p. 16.

10 'Escalation dominance' refers to a situation whereby an actor escalates the conflict in a manner that will be disadvantageous to an adversary; the adversary is unable, however, to respond in kind – either because it has limited escalation options, or those options would not alter its predicament. Escalation dominance is thought by some to be unobtainable in a world of multipolar nuclear powers. See Morgan et al., *Dangerous Thresholds: Managing Escalation in the 21st Century*, p. 15; Charles L. Glaser, *Analyzing Strategic Nuclear Policy* (Princeton, NJ: Princeton University Press, 1990), pp. 50–57; and Herman Kahn, *On Escalation: Metaphors and Scenarios* (New York: Praeger, 1965), p. 23.

11 For example, analysts have expressed concerns that China's ballistic missiles, armed with both nuclear and conventional variants (e.g. DF-21 and DF-26), increase the risk that the US might mistake a conventionally armed missile for a nuclear one, in turn, triggering an inadvertent escalation. James Johnson, "China's Guam Express and Carrier Killers: The Anti-Ship Asymmetric Challenge to the US in the Western Pacific," *Comparative Strategy*, 36, 4 (2017), pp. 319–332.

12 Robert Jervis, *The Illogic of American Nuclear Strategy* (Ithaca, NY: Cornell University Press, 1984).

13 During the Cold War, escalation management became almost synonymous with preventing conventional wars from escalating to nuclear conflict and managing the scope and intensity of nuclear warfighting after initial nuclear use. Richard Smoke, *War: Controlling Escalation* (Cambridge, MA: Harvard University Press, 1977).

14 The broader notion that escalation can occur inadvertently (or unintentionally) is attributed to a failure to appreciate the pressures that one's actions put on an adversary. These pressures generate perceived first-mover advantages and are associated with the 'security dilemma' concept – when defensively motivated actions can appear and be perceived as offensive. Robert Jervis, "Cooperation under the Security Dilemma," *World Politics*, 30, 2 (1978), pp. 167–214.

15 Ibid.

16 For recent analysis on the use of emerging technologies for signaling purposes, see Evan B. Montgomery, "Signals of Strength: Capability Demonstrations and Perceptions of Military Power," *Journal of Strategic Studies*, 43, 2 (2019), pp. 309–330.

17 Emerging technology could also merely enable the intentional escalation behavior that states would have undertaken in some other form, even in the absence of a particular technology. Caitlin Talmadge, "Emerging Technology and Intra-War Escalation Risks: Evidence from the Cold War, Implications for Today," *Journal of Strategic Studies*, 42, 6 (2019)," pp. 864–887, p. 883.

18 See Thomas J. Christensen, "The Meaning of the Nuclear Evolution: China's Strategic Modernization and US–China Security Relations," *Journal of Strategic Studies*, 35, 4 (2012), pp. 447–487; and Pavel Podvig, "Russia and the Prompt Global Strike Plan," *PONARS Policy Memo*, No. 417, December 2006.

19 Ben Connable, *Embracing the Fog of War: Assessment and Metrics in Counterinsurgency* (Santa Monica, CA: RAND Corporation, 2012).
20 The distinction between the *actual* risk of escalation and *perceived* risk is largely irrelevant in determining whether the resultant effect is stabilizing or destabilizing. The effect of fear on escalation risk depends in large part on perceptions, not actual risk. Miles, "The Dynamics of Strategic Stability and Instability," p. 437.
21 Patricia Lewis and Beyza Unal, *Cybersecurity of Nuclear Weapons Systems: Threats, Vulnerabilities and Consequences* (London: Chatham House Report, Royal Institute of International Affairs, 2018), p. 16.
22 'Asymmetric information' refers to a situation where one state has information that is different from what the other states know. Imperfect (or asymmetric) knowledge of the other side alone is not sufficient to cause war; in theory, adversaries should communicate this information to each other.
23 Deterrence theorists describe the trade-offs that exist at the level at which to calibrate states' retaliation threshold, how quickly to gradate it, and how explicitly to declare it to an adversary. An asymmetric escalation deterrence posture, in contrast to an assured retaliatory one, is considered to produce more deterrence, but entail higher escalatory risks. See Vipin Narang, *Nuclear Strategy in the Modern Era: Regional Powers and International Conflict* (Princeton, NJ: Princeton University Press, 2014), pp. 17–20.
24 See James Johnson, "Chinese Evolving Approaches to Nuclear 'Warfighting': An Emerging Intense US–China Security Dilemma and Threats to Crisis Stability in the Asia Pacific," *Asian Security*, 15, 3 (2019), pp. 215–232.
25 For example, see Lin Juanjuan, Zhang Yuantao, and Wang Wei "Military Intelligence is Profoundly Affecting Future Operations," *Ministry of National Defense of the People's Republic of China*, September 10, 2019, www.mod.gov.cn/jmsd/2019-09/10/content_4850148.htm (accessed October 10, 2019).
26 Yuna Huh Wong et al., *Deterrence in the Age of Thinking Machines* (Santa Monica, CA: RAND Corporation, 2020), p. 53.
27 Ibid., p. 68.
28 James D. Fearon, "Rationalist Explanations for War," *International Organization*, 49, 3 (1995), pp. 379–414.
29 An alternative notion is that inaccurate, conflicting, or ambiguous information can cause actors to perceive (or exaggerate) things that do not exist, and thus be deterred from taking escalatory action – or 'self-deterrence.' Robert Jervis *How Statesmen Think: The Psychology of International Politics* (Princeton, NJ: Princeton University Press, 2017), pp. 201–202.
30 Brendan R. Green and Austin Long, "Conceal or Reveal? Managing Clandestine Military Capabilities in Peacetime Competition," *International Security*, 44, 3 (Winter 2019/20), pp. 50–51.
31 The literature on uncertain military capabilities and warfare demonstrates a clear link to bargaining failure and conflict. See Mark Fey and Kristopher W. Ramsay, "Uncertainty and Incentives in Crisis Bargaining: Game-Free Analysis of International Conflict," *American Journal of Political Science*,

55, 1 (January 2011), pp. 149–169; William Reed, "Information, Power, and War," *American Political Science Review*, 97, 4 (November 2003), pp. 633–641; and Fearon, "Rationalist Explanations for War."
32 Conversely, prospect theory posits that in such situations where the military balance is unclear, both sides will be more risk-averse. Opacity about the balance could, therefore, dissuade an actor from considering military force during a crisis. Green and Long, "Conceal or Reveal? Managing Clandestine Military Capabilities in Peacetime Competition," pp. 48–83.
33 Ibid., pp. 50–51.
34 On the centrality of uncertainty in the use of military force and the advent of war see Erik Gartzke, "War is in the Error Term," *International Organization*, 53, 3 (1999), pp. 567–587.
35 Stephen D. Biddle, and Robert Zirkle, "Technology, Civil–Military Relations, and Warfare in the Developing World," *Journal of Strategic Studies*, 19, 2 (1996), pp. 171–212.
36 For example, US counterforce capabilities developed during the Cold War came about for fundamentally non-technological reasons – Congressional funding priorities to support an ambitious nuclear force posture. Austin Long and Brendan Rittenhouse Green, "Stalking the Secure Second Strike: Intelligence, Counterforce, and Nuclear Strategy," *Journal of Strategic Studies*, 38, 1–2 (2014), pp. 38–73.
37 Talmadge, "Emerging Technology and Intra-War Escalation Risks."
38 The US strategic community has expressed concern that the AirSea Battle Concept (re-named Joint Concept for Access and Maneuver in the Global Commons) will increase the risk of nuclear escalation. See Jan Van Tol et al., *AirSea Battle: A Point-of-Departure Operational Concept* (Washington, DC: Center for Strategic and Budgetary Assessments, May 18, 2010).
39 The evidence does not suggest that China intends to use nuclear weapons to achieve escalation dominance for deliberate escalation – or nuclear warfighting – to compel the US and its allies to back down during crisis or conflict. Johnson, "Chinese Evolving Approaches to Nuclear 'Warfighting'."
40 For Chinese views on escalation see Christensen, "The Meaning of the Nuclear Evolution;" Fiona Cunningham and M. Taylor Fravel, "Assuring Assured Retaliation: China's Nuclear Posture and US–China Strategic Stability," *International Security*, 40, 2 (2015), pp. 40–45; Morgan et al., *Dangerous Thresholds: Managing Escalation in the 21st Century*, chapter 3; and Avery Goldstein, "First Things First: The Pressing Danger of Crisis Instability in US–China Relations," *International Security*, 37, 4 (Spring 2013), pp. 67–68.
41 Morgan et al., *Dangerous Thresholds: Managing Escalation in the 21st Century*, pp. 169–170.
42 The Chinese conceptualization of 'strategic deterrence' includes nuclear deterrence and conventional deterrence and information deterrence. Chinese military writing often conflates the concept alongside the 'people's warfare' strategic concept dating back to the Mao era. See Xiao Tianliang (ed.), *The Science of*

*Military Strategy* (Beijing: National Defense University Press, 2015/201); and Jeffrey E. Kline and Wayne P. Hughes Jr., "Between Peace and the Air-Sea Battle: A War at Sea Strategy," *Naval War College Review*, 65, 4 (Autumn 2012), pp. 34–41.

43 See James M. Acton (ed.), with Li Bin, Alexey Arbatov, Petr Topychkanov, and Zhao Tong, *Entanglement: Russian and Chinese Perspectives on Non-Nuclear Weapons and Nuclear Risks* (Washington, DC: Carnegie Endowment for International Peace, 2017).

44 This kind of deterrence logic underpinned the Cold War-era 'tripwire' military deployments by NATO's nuclear-armed states to deter the Soviet Union. See Thomas C. Schelling, *Arms and Influence* (New Haven, CT: Yale University Press, 1966), p. 47.

45 Before the 1990s, states' nuclear and conventional support systems (i.e. communications, early warning, intelligence, and battlefield situational awareness) primarily functioned independently of one another. As a result, these systems came to be viewed as contributing positively to strategic stability by ensuring confidence in the durability of the overall nuclear deterrent. By contrast, modern support systems are increasingly co-mingled (or dual-use), with blurred boundaries between conventional and nuclear support systems.

46 Joshua Rovner, "Two Kinds of Catastrophe: Nuclear Escalation and Protracted war in Asia," *Journal of Strategic Studies*, 40, 5 (2017), pp. 696–730.

47 For example, decision-makers may develop and deploy technologies to enable them to manipulate escalatory risk in new ways (i.e. to achieve escalation dominance), or to opportunistically use technology to fulfill deliberately escalatory operations, which they want to pursue for other reasons. Talmadge, "Emerging Technology and Intra-War Escalation Risks," p. 869; and Keir Lieber, "Grasping the Technological Peace: The Offense–Defense Balance and International Security," *International Security*, 25, 1 (2000), pp. 71–104.

48 Talmadge, "Emerging Technology and Intra-War Escalation Risks," p. 882.

49 Conversely, reduced human casualties remove the requirement for personnel recovery missions and, thus, reduce pressure for retaliation – though the system's value or sensitivity may warrant action to recover or destroy it, nonetheless.

50 Acton (ed.), with Bin, Arbatov, Topychkanov, and Tong, *Entanglement: Russian and Chinese Perspectives on Non-Nuclear Weapons and Nuclear Risks*, p. 48.

51 For example, during the Cold War, it would have been unthinkable for the Soviet Union to suggest that a conventional war in Europe would leave Moscow vulnerable to a nuclear first strike, which US observers frequently highlighted.

52 On this idea see Daniel Byman and Matthew Waxman, *The Dynamics of Coercion: American Foreign Policy and the Limits of Military Might* (Cambridge: Cambridge University Press, 2002), Chapter 8.

53 Thomas C. Schelling, *The Strategy of Conflict* (Cambridge, MA: Harvard University Press, 1960), chapter 8, p. 187.

54 'Escalation ladder' in this context refers to the forty-four 'rungs' on a metaphorical ladder of escalating military conflict. For the seminal text that introduced

the concepts of an 'escalation ladder' as it applies to the entire range of conflict from conventional conflict to all-out nuclear warfare see Herman Kahn, *On Escalation: Metaphors and Scenarios* (New York: Praeger, 1965).

55 The main types of escalation mechanisms include deliberate, inadvertent, unauthorized, and accidental escalation. There is also a distinction between vertical and horizontal escalation. These categorizations are not, however, hard and fast; for example, accidental escalation can also be inherently unauthorized. See Morgan et al., *Dangerous Thresholds: Managing Escalation in the 21st Century*, pp. 18–20. Much of the nuclear escalation literature has focused on the conditions that could lead to unauthorized nuclear use or accidental launch based on false warnings. For example, see Bruce Blair, *Strategic Command and Control: Redefining the Nuclear Threat* (Washington, DC: The Brookings Institution, 1989); Scott Sagan, *The Limits of Safety: Organizations, Accidents, and Nuclear Weapons* (Princeton, NJ: Princeton University Press, 1995); and Peter Feaver, *Guarding the Guardians: Civilian Control of Nuclear Weapons in the United States* (Ithaca, NY: Cornell University Press 1992).

56 For example, during the Korean War, the US failed to recognize that if the US conquered North Korea and continued to march northwards, it would threaten China; thus, deterrence failed. The historical record reveals that Beijing attempted to communicate its intention to enter the war – explicitly warning, by way of India, the consequences of the US crossing the 38th parallel – but the Chinese did not realize that the situation was not understood in Washington. See Harvey DeWeerd, "Strategic Surprise in the Korean War," *Orbis*, 6, Fall (1962), pp. 435–452.

57 Not all escalation results from decisions, however. For example, indecision, accidents, procedures, and protocols can all escalate situations *without* any actual decision-making having taken place.

58 Lawrence Lewis and Anna Williams, *Impact of Unmanned Systems to Escalate Dynamics (Summary)* (Washington, DC: Center for Naval Analysis, 2018), www.cna.org/CNA_files/PDF/Summary-Impact-of-Unmanned-Systems-to-Escalation-Dynamics.pdf (accessed August 10, 2019).

59 When two states have conflicting views about when the firebreaks during a crisis or conflict occur, each side might focus on negotiating an end to the situation at different times, thus forfeiting the opportunity to de-escalate the situation. See Bernard Brodie, *Escalation and the Nuclear Option* (Princeton, NJ: Princeton University Press, 1966).

60 Chris Buckley, "Chinese Navy Returns Seized Underwater Drone to the US," *New York Times*, December 12, 2016, www.nytimes.com/2016/12/20/world/asia/china-returns-us-drone.html (accessed August 10, 2019).

61 For example, in 2013, Pakistan publicly denounced US drones violating its sovereign airspace. Similarly, in 2019, Iran shot down an unmanned US RQ-4 *Global Hawk* unmanned surveillance aircraft, accusing the US of operating military operations in its airspace disputed by US officials. Richard Leiby, "U.N.: US Drone Strikes Violate Pakistan Sovereignty," *Washington Post*, March 15, 2013, www.washingtonpost.com/world/asia_pacific/un-us-dron

es-violate-pakistan-sovereignty/2013/03/15/308adae6-8d8a-11e2-adca-74ab31da3399_story.html (accessed August 10, 2019); and Jim Garamone, "Iran Shoots Down US Global Hawk Operating in International Airspace," US Department of Defense, June 20, 2019, www.defense.gov/Explore/News/Article/Article/1882497/iran-shoots-down-us-global-hawk-operating-in-international-airspace/ (accessed August 10, 2019).

62 See James S. Johnson, "Artificial Intelligence: A Threat to Strategic Stability," *Strategic Studies Quarterly*, 14, 1 (2020), pp. 16–39.

63 M. Taylor Fravel, *Active Defense China's Military Strategy since 1949* (Princeton, NJ: Princeton University Press, 2019).

64 James S. Johnson, "Chinese Nuclear 'War-fighting:' "An Emerging Intense US–China Security Dilemma and Threats to Crisis Stability in the Asia Pacific," *Asian Security*, 15, 3 (2019), pp. 215–232.

65 Avery Goldstein, "First Things First: The Pressing Danger of Crisis Instability in US–China Relations," pp. 65–66; and James Johnson, "Artificial Intelligence in Nuclear Warfare: A Perfect Storm of Instability?" *The Washington Quarterly*, 43, 2 (2020) pp. 197–211.

66 Fiona S. Cunningham and M. Taylor Fravel, "Dangerous Confidence? Chinese Views on Nuclear Escalation," *International Security* (2019), 44, 2, pp. 106–108.

67 During the Cold War, the 'slippery slope' from conventional to nuclear war was created by the lack of a US NFU policy, combined with deployments of tactical nuclear weapons designed to offset Soviet conventional numerical superiority in Central Europe. Robert Jervis, *The Meaning of the Nuclear Revolution: Statecraft and the Prospect of Armageddon* (Ithaca, NY: Cornell University Press, 1989).

68 International relations scholars generally agree that rational incentives for states to exaggerate (or understate) their resolve can cause misperceptions about resolve – in this case, the threshold for the use of nuclear weapons increases the risk of escalation. Fearon, "Rationalist Explanations for War," pp. 393–395. These concerns were raised in the early days of the Cold War when debates arose about how best to fight conventional wars beneath the nuclear threshold, given the paramount goal of avoiding major nuclear war. See Robert Powell, "Bargaining Theory and International Conflict," *Annual Review of Political Science*, 5, 1 (2001), pp. 1–30; and Posen, *Inadvertent Escalation: Conventional War and Nuclear Risks*.

69 James M. Acton, *Silver Bullet? Asking the Right Questions About Conventional Prompt Global Strike* (Washington, DC: Carnegie Endowment for International Peace, 2013), p. 71.

70 Chinese confidence in its ability to control nuclear escalation can be explained by: (1) China's long-standing decoupling of conventional and nuclear strategy; (2) the availability of non-nuclear strategic weapons (e.g. cyber weapons); (3) a military organizational bias that tends to favor nuclear experts and the PLA's Rocket Force leadership; and (4) a belief that the US would avoid intervening in a conflict between a US ally and China if doing so risked nuclear confrontation.

Avery Goldstein, "First Things First: The Pressing Danger of Crisis Instability in US–China Relations," pp. 49–89.

71 For example, China's strategic community does not appear to share US concerns that the use of conventional weapons to implement the AirSea Battle Concept would increase the risk of inadvertent nuclear escalation. See Johnson, "Chinese Nuclear 'War-fighting;" and Jeffrey A. Larsen and Kerry M. Kartchner (eds), *On Limited Nuclear War in the 21st Century* (Palo Alto, CA: Stanford University Press, 2014).

72 Conversely, Beijing may anticipate US efforts to escalate a conventional war to achieve conventional dominance, and Washington may anticipate a disproportionate Chinese nuclear response if it were to conduct limited nuclear strikes in conflict, thereby encouraging caution on both sides. Cunningham and Fravel, "Dangerous Confidence? Chinese Views on Nuclear Escalation."

73 It is noteworthy that the 2018 US *Nuclear Posture Review* asserts that Beijing might believe that it could secure advantages through the limited use of nuclear weapons. US Department of Defense, *Nuclear Posture Review* (Washington, DC: US Department of Defense, February 2018), p. 32.

74 Cunningham and Fravel, "Dangerous Confidence? Chinese Views on Nuclear Escalation," p. 104–105.

75 The 2015 US DoD Cyber Strategy explicitly identifies an adversary's command and control systems as a target for US offensive cyber operations. US Department of Defense, *The DoD Cyber Strategy* (Washington DC: April 2015), https://archive.defense.gov/home/features/2015/0415_cyberstrategy/final_2015_dod_cyber_strategy_for_web.pdf, p. 14 (accessed March 10, 2020).

76 See Lora Saalman, "The China factor," in Alexei Arbatov, Vladimir Dvorkin, and Natalia Bubnova (eds), *Missile Defense: Confrontation and Cooperation* (Moscow: Carnegie Moscow Center, 2013), pp. 226–252; and Acton, *Silver Bullet? Asking the Right Questions About Conventional Prompt Global Strike.*

77 It remains unclear whether Chinese analysts appreciate the inadvertent and accidental escalation risks associated with the PLA's offensive *conventional* missile and space-warfare doctrine. Morgan et al., *Dangerous Thresholds: Managing Escalation in the 21st Century*, p. 78.

78 It is believed that China's C3I systems support both its nuclear and non-nuclear forces. Similarly, Russian strategic submarines and bombers are located on the same bases as general-purpose naval vessels and aircraft.

79 Cunningham and Fravel, "Assuring Assured Retaliation: China's Nuclear Posture and US–China Strategic Stability," p. 10.

80 The extent of the PLA's nuclear and conventional co-mingling is challenging to collaborate with existing open-source materials. While China appears to operate geographically and organizationally distinct launch brigades for its nuclear and conventional missiles, deployment locations for nuclear and non-nuclear variants of the same missile type may overlap, potentially resulting in co-mingling, and thus increasing the escalation risks associated with the ambiguity and uncertainty caused by this overlap. Were China to use co-mingling to manipulate the threat of nuclear escalation, however, the extent of

co-mingling would likely be much greater than current capability assessments suggest. Caitlin Talmadge, "Would China Go Nuclear? Assessing the Risk of Chinese Nuclear Escalation in a Conventional War with the United States," *International Security*, 41, 4 (Spring 2017), pp. 50–92.
81 The US Department of Defense, *Military and Security Developments Involving the People's Republic of China, 2019* (Washington, DC: US Department of Defense, May 2019) https://media.defense.gov/2019/May/02/2002127082/-1/-1/1/2019_CHINA_MILITARY_POWER_REPORT.pdf (accessed March 10, 2020), p. 66.
82 To date, open sources indicate that no consensus exists within China's strategic community on the importance of separating nuclear and conventional forces. Zhao, Tong, "Conventional challenges to strategic stability: Chinese perceptions of hypersonic technology and the security dilemma," in Lawrence Rubin and Adam N. Stulberg (eds), *The End of Strategic Stability? Nuclear Weapons and the Challenge of Regional Rivalries* (Washington, DC: Georgetown University Press, 2018), pp. 174–202, p. 195.
83 Li Bin, "Tracking Chinese Strategic Mobile Missiles," *Science and Global Security*, 15 (2007), pp. 1–30; Charles L. Glaser and Steve Fetter, "Should the United States Reject MAD? Damage Limitation and US Nuclear Strategy toward China," *International Security*, 41, 1 (2016), pp. 49–98.
84 Cunningham and Fravel, "Assuring Assured Retaliation: China's Nuclear Posture and US–China Strategic Stability," pp. 19–20.
85 For example, over the past two decades, US satellites have developed techniques to detect moving targets, determine their speed and direction, and locate mobile missiles, more effectively than in the past. Keir A. Lieber and Daryl G. Press, "Why States Won't Give Nuclear Weapons to Terrorists," *International Security*, 38, 1 (Summer 2013), pp. 80–104, p. 38.
86 "Why Intelligent Operations Will Win in Future Warfare," *Xinhua News Service*, January 15, 2019, www.xinhuanet.com/mil/2019–01/15/c_1210038327.htm (accessed August 10, 2019).
87 To date, Chinese analysts have not studied in-depth studies on the implications of deploying advanced military technologies for escalation. Moreover, the stove piping inherent within China's military constrains discussion within China's strategic community and hinders meaningful exchanges with foreign defense communities. Yan Guoqun and Tao Zhonghua, "New weapon in the world: The development of hypersonic weapons draws attention" *PLA Daily*, February 19, 2003, www.people.com.cn/GB/junshi/63/20030219/925824.html (accessed August 10, 2019).
88 A crucial lesson from the Cuban Missile Crisis is that standard operating procedures are often poorly understood by leaders. Escalatory signals may be sent even when escalatory intent is absent.
89 'Escalation management' is about keeping military confrontations from moving up the rungs to all-out war and ensuring that limited wars are controlled. Because escalation is an interactive phenomenon, involving fluid thresholds that shift throughout a conflict, it can rarely be *controlled* in the ordinary sense

of the word. Morgan et al., *Dangerous Thresholds: Managing Escalation in the 21st Century*, p. 160.

90 The escalation pathways progress a crisis or conflict to all. Our warfare generally involves several escalation mechanisms; for example, an accidental escalation situation might trigger a deliberate escalatory response. Ibid., p. 28.
91 Escalatory actions could also be taken to signal to an adversary about an escalation that may or may not occur in the future – or 'suggestive escalation.' Thomas C. Schelling, *Arms and Influence* (New Haven, CT: Yale University Press, 1966).
92 The importance of empathizing with one's enemy has been long extolled by scholars of deterrence theory as central to managing inadvertent and accidental escalation risks. See Robert Jervis, *Perception and Misperception in International Politics* (Princeton, NJ: Princeton University Press, 1976), chapter 3.
93 David Danks, "Singular causation," in M.R. Waldmann (ed.), *Oxford Handbook of Causal Reasoning* (Oxford: Oxford University Press, 2017), pp. 201–215.
94 The US is concerned about rising powers such as Russia and China. Russia has contingencies for confrontations with both the US and China; China considers Russia, India, and above all the US, as potential adversaries; India is embroiled in strategic competition with both China and Pakistan; and finally, North Korea is a strategic concern for the other nuclear-weapon powers. Geist and Lohn, *How Might Artificial Intelligence Affect the Risk of Nuclear War?*, p. 12.
95 The White House, *National Security Strategy of the United States of America*, December 2017, www.whitehouse.gov/wp-content/uploads/2017/12/NSS-Final-12-18-2017-0905-2.pdf (accessed June 10, 2019).
96 John J. Mearsheimer, *Conventional Deterrence* (Ithaca, NY: Cornell University Press, 1984), p. 210.
97 Glenn Snyder, *Deterrence and Defense* (Princeton, NJ: Princeton University Press, 1961).
98 Christensen, "The Meaning of the Nuclear Evolution: China's Strategic Modernization and US–China Security Relations," pp. 467–471; Acton, *Silver Bullet? Asking the Right Questions About Conventional Prompt Global Strike*; and Talmadge, "Would China Go Nuclear?" pp. 50–92.
99 Keren Yarhi-Milo, *Knowing the Adversary* (Princeton, NJ: Princeton University Press, 2014), p. 250.
100 Johnson, "Artificial Intelligence in Nuclear Warfare," p. 207.
101 Political psychologists have demonstrated that states tend to view an adversary's behavior as more coordinated, centralized, and coherent than it is. Thus, actions that are caused by decisions made at different parts of a decentralized bureaucracy, or that are the result of accidental or inadvertent behavior, will likely be interpreted as representative of a broader and potentially nefarious plot. See Jervis, *How Statesmen Think*, p. 225.

# Part III

Nuclear instability redux?

# 5

# Hunting for nuclear weapons in the digital age

Part III of the book includes four case studies to elucidate the escalation risks associated with AI. These studies demonstrate how and why military AI systems fused with advanced strategic non-nuclear weapons (or conventional counterforce capabilities) might cause or exacerbate escalation risks in future warfare.[1] They also illuminate how these AI-augmented capabilities would work; despite the risks associated with their deployment, great military powers will likely deploy them. Military commanders obsessed with tightly controlling the rungs on the escalation ladder should, in theory, be against delegating too much decision-making authority to machines, especially those involving nuclear weapons. Chapters 5 to 8 find that competitive pressures between great military powers, in particular the fear that others will gain the upper hand in AI – and the capabilities AI empowers – will likely create pressure that causes states to eschew these concerns.

How might AI-augmented intelligence gathering, and analysis systems impact the survivability and credibility of states' nuclear-deterrent forces? Technologies such as AI, machine learning (ML), and big-data analytics associated with the 'computer revolution' have the potential to significantly improve the ability of militaries to locate, track, target, and destroy a rival's nuclear-deterrent forces without the need to deploy nuclear weapons.[2] Thus, AI applications that make survivable strategic forces, such as submarines and mobile missiles, more vulnerable (or perceived as such), could have destabilizing escalatory effects, even if the state in possession of these counterforce capabilities did not intend to use them.

As strategist Alfred Mahan famously observed, "force is never more operative than when it is known to exist but is not brandished."[3] Advances in deep learning, for example, can exponentially improve machine vision and other signal processing applications, which may overcome the main technical barriers for tracking and targeting adversaries' nuclear forces (i.e. sensing, image processing, and estimating weapon velocities and kill radius). Some scholars argue that AI and autonomy could enable real-time tracking and more accurate targeting of an adversary's nuclear assets in ways that make counterforce operations more feasible.[4]

Moreover, the speed of AI could put the defender at a distinct disadvantage, creating additional incentives to strike first (or pre-emptively) against technologically superior military rivals. Consequently, the less secure a nation considers its second-strike capabilities to be, the more likely it is to countenance the use of autonomous systems within its nuclear weapons complex to bolster the survivability of its strategic forces. According to analyst Paul Scharre, "winning in swarm combat may depend upon having the best algorithms to enable better coordination and *faster reaction times*, rather than simply the best platforms."[5]

Advances in AI ML techniques could significantly improve existing machine vision and other signal processing applications, and enhance autonomy, and sensor fusion applications. Strengthening the functionality of intelligence, surveillance, and reconnaissance (ISR), automatic target recognition (ATR), and terminal guidance systems would have profound implications for strategic stability.[6] Moreover, AI used in conjunction with autonomous mobile sensor platforms might compound the threat posed to the survivability of mobile intercontinental ballistic missiles (ICBM) launchers.[7] Autonomous mobile sensors would need to locate close to mobile ICBM launchers to be effective, and thus, as the 'window of vulnerability' rapidly narrowed and faced with the prospect of an imminent disarming strike, an adversary would be put under immense pressure to escalate.

The remainder of this chapter proceeds in two sections. The first identifies a range of AI-enhanced intelligence analysis systems (i.e. ISR sensors, ATR, terminal guidance capabilities, and data-analysis systems) used in conjunction with other advanced capabilities, which could significantly enhance the ability of militaries to locate land-based nuclear mobile missiles.[8] It argues that in combination, enhanced ISR systems and more robust missile defenses may create incentives for states to use pre-emptive strikes against an adversary's strategic assets.

The second section examines how AI-augmented anti-submarine warfare (ASW) – especially big-data analytics and ML – might further reduce the deterrence utility of nuclear-powered ballistic missile submarines (SSBNs). It argues that AI could soon overcome some of the remaining technical barriers to reliably and accurately locate and track submarines, thereb, eroding the deterrence utility of stealthy SSBNs and making use-them-or-lose-them situations more likely to occur. The technical feasibility of this hypothesis remains, however, highly contested.

## Hunting for strategic mobile missiles

AI enhancements could significantly improve the accuracy, speed, and reliability of ISR, sensor technology, ATR, terminal guidance capabilities, and data-analysis systems. For instance, drone swarms deployed in nuclear-ISR missions – and supported by ML technology – may enhance sensor drones' reliability and speed to evade enemy defenses and locate mobile missiles. Theoretically, satellite imagery and signals intelligence from drone swarms in real-time would cue stealth fighters, or sorties of armed drones, to destroy these missiles.[9]

Because of the inherent difficulty in finding mobile missiles, even modest improvements towards enabling this capability (or even the perception of vulnerability) could be a strategic game-changer.[10] Moreover, the speed at which AI-enhanced ISR systems might target and execute kinetic operations could limit de-escalation options. Several technologies under development are designed explicitly for this purpose. For example, the US Navy is currently testing an autonomous surface trail vehicle prototype with a short-range sonar (*Sea Hunter*).[11] To date, these technologies are not sufficiently mature to pose a credible threat to states' confidence in their second-strike capabilities.[12] Besides, the technical feasibility of this hypothesis remains highly contested.[13] In the near-term, therefore, because of these technical challenges, nuclear deterrence based on mutually assured destruction is unlikely to be upended by AI-augmented counterforce capabilities.[14]

Some experts on China's nuclear forces argue that it is highly improbable the US would be able to develop the capability to locate all of China's land-based mobile missiles, as long as China maintains rudimentary deception and dispersal tactics.[15] Were this confidence to be undermined by technical improvements in the accuracy and remote sensing of US counterforce capabilities, however, the implications for the survivability of Chinese mobile missiles and, in turn, crisis stability would be profound. As shown in chapter 4, Chinese and US overconfidence in ability to control escalation would exacerbate the risk that a conventional war could inadvertently escalate to the nuclear level.[16] If Chinese leaders are fearful that an imminent US counterforce attack might circumvent its deception and dispersal tactics, and upend its survivable nuclear arsenals as a prelude to a disarming first strike, Beijing may feel compelled (and despite its no-first-use pledge) to engage in limited nuclear escalation to reclaim the upper hand – or escalation dominance.[17]

Given the tendency of Chinese (and Russian) strategists to extrapolate from current US capabilities malign intent, and to assume future ones will threaten their security, even modest and incremental improvements in AI techniques to integrate and synthesize data about the location of an adversary's mobile missiles could exacerbate pre-existing fears and distrust.[18] Irrespective of whether future breakthroughs in AI produce irrefutable evidence of a game-changing means of locating, targeting, and destroying mobile missile forces, Chinese and Russian perceptions of US intentions in the pursuit of these capabilities would, therefore, be far more salient. In other words, despite the reassurances of the US, its adversaries would be unable to dismiss the possibility that military AI capabilities would not be used in future warfare to erode the survivability of their nuclear forces – a contingency the US has prepared for over several decades.[19]

Furthermore, the challenge of determining an attacker's intentions would be complicated if an adversary's dual-use ISR, early warning, command, control, communications, and intelligence (C3I) systems were targeted earlier in future warfare.[20] In a future conflict between the US and China (or the US and Russia), both sides would have strong incentives to attack the enemy's dual-use C3I capabilities early on and pre-emptively.[21] Effective deterrence depends on the clear communication of credible threats (and consequence of violation) between adversaries, which assumes the sender and recipient of these signals share a common context allowing for mutual interpretation.[22] Both sides would likely assume the worst and respond accordingly, denied critical real-time battlefield information, and compounded by the fog of war.[23] Any increase in the vulnerability of states' nuclear mobile systems caused by AI innovations might upend the deterrence assumptions that undergirded the nuclear revolution – relying on hardening and concealing approaches for ensuring the survivability and credibility of their nuclear forces.[24]

Theoretically, a state could launch long-range conventional missile salvos in conjunction with big-data analytics, cyber capabilities, and AI-enabled autonomous weapons. Then, it could use its missile defences, supported by AI-augmented ISR, ATR systems, to mop up any remaining nuclear retaliatory capabilities.[25] In combination, enhanced ISR systems and sophisticated missile defenses could increase the incentives for pre-emptive strikes targeting an adversary's strategic assets (i.e. mobile missile and dual-use early warning and C3I systems.[26] As shown in chapter 4, this kind of interoperability, coupled with states' increasingly co-mingled nuclear and non-nuclear capabilities, might create a *de facto* escalation framework for nuclear deterrence.[27]

While threats below the nuclear level would be politically more acceptable, it might force an adversary to make a Hobson's choice between capitulation and escalation with nuclear weapons.[28] According to analyst Eric Heginbotham, "the hunt for conventionally armed missiles could result in the attrition of China's nuclear-capable missile force," which could create a destabilizing use-them-or-lose-them dilemma.[29] In sum, the capabilities AI might *enhance* (cyber-weapons, drones, precision strike missiles, and hypersonic weapons), together with the ones it might *enable* (ISR, ATR, and autonomous sensor platforms) could make hunting for mobile nuclear arsenals faster, cheaper, and more effective than before.[30] In aggregate, AI-enabled and enhanced capabilities will likely have a more significant impact on strategic stability than the sum of its parts. Put another way, the strategic implications of adopting military AI will likely be greater than those of any specific military task.

## AI-enabled 'ocean transparency'?[31]

In the maritime domain (unmanned undersea vehicles (UUVs), unmanned surface vehicles (USVs), and unmanned aerial vehicles (UAVs)), supported by AI-enhanced intra-swarm communication and ISR systems, may be simultaneously deployed in *both* offensive and defensive anti-submarine warfare operations in order to saturate an enemy's defenses and locate, disable, and destroy its nuclear-armed or non-nuclear attack submarines.[32] Because of the stealth technology (notably minimal acoustic signatures) of modern diesel-electric powered attack submarines and SSBNs, coupled with the immense challenge of coordinating such an operation, tracking a submarine from a surface ship (or even from another submarine) is, even in relatively benign conditions, a challenging proposition.[33]

Despite continued advances in sensor technology design to overcome challenges of submarine quieting in anti-submarine warfare (i.e. reduced size and extended detection ranges), other technical challenges remain. These include underwater communication between multiple systems, processing power requirements, battery life, and scaling the system to blue water.[34] However, none of these technical hurdles appear insurmountable, and some of the physical limitations could be mitigated by swarming.[35] Therefore, instead of making submarines redundant, modern ASW capabilities have reduced the effectiveness of this capability, slowing the deployment of submarines in patrol areas, inhibiting them from getting into firing position, and disrupting the coordination of attacks.[36]

According to scholar Owen Cote, "the question of whether submarines are getting harder to hide depends very much on whose submarines you're talking about, who's hunting them, and where."[37] For example, by operating in an area in which several countries are using submarines, SSBNs could potentially remain undetected. That is, the larger the number of submarines – both strategic and non-strategic – operating at sea, the lower the chances that a space-based or high-altitude ocean surveillance system could reliably identify strategic submarines.

Recent advances in sensor, communication, and processing technologies – especially big-data analytics and ML – could become disruptive, transformative technologies in future ASW and undersea support platforms (e.g. UUVs, USVs, and UAVs), to locate and attack submarines in real-time, and enhance the stealth and endurance of submarines and their attendant weapon systems. A combination of AI ML and big-data analytics might allow Cold War-era sensitivity technology to detect radiation and chemical emissions from submarines, enabling new capabilities to locate and cue torpedo seekers in long-range anti-submarine 'fire and forget' operations.[38]

Several observers posit that autonomous systems like the Defense Advanced Research Projects Agency (DARPA)'s *Sea Hunter* by rendering the underwater domain 'transparent,' might erode the second-strike deterrence utility of stealthy SSBNs, triggering use-them-or-lose-them situations. Others do not expect a technically reliable and effective capability of this kind will be operational for the foreseeable future.[39] For now, however, the technical feasibility of this hypothesis remains highly contested. On the one hand, several experts posit that emerging technologies such as AI, quantum communications, and big-data analytics will empower new iterations of highly portable sensing, communications, and signal-processing platforms that could render at-sea nuclear deterrence all but obsolete.[40]

On the other hand, some consider this hypothesis technically and operationally premature for the following reasons. First, it is unlikely that sensors onboard ASW would be able to detect deeply submerged submarines reliably. Second, ASW long-range sensors would require a reliable, robust multi-system control system for optimal area surveillance, tracking, and data transmission – a capability that has not been developed. Third, the range of these sensors (and the drones themselves) would be limited by battery power over extended ranges.[41] Finally, and related, given the vast areas traversed by SSBNs on deterrence missions, the chance of detection is negligible, even if a large number of autonomous swarms were deployed on reconnaissance missions.[42]

Furthermore, in notoriously complex and dynamic underwater conditions the transit time required for armed ASW to reach their target would leave a wide enough window for the SSBNs to conduct tit-for-tat countermeasures, such as decoys and maneuver tactics, jamming, deception (e.g. deploying UUVs designed to mimic tonals frequencies of SSBNs), or destroying ASW sensors with non-kinetic cyber or electronic warfare capabilities.[43] As shown in chapter 2, the Cold War experience demonstrated the Neophiliac dangers of assuming that disruptive technologies will necessarily have game-changing strategic effects.[44]

Significant advances in power, sensor technology, and communications would be needed before autonomous systems have a game-changing strategic impact on submarine reconnaissance.[45] Even without these improvements, AI-enabled computer processing, real-time oceanographic modeling techniques, and drone swarming capabilities (i.e. UAVs, UUVs, and USVs) will likely nonetheless have a significant qualitative impact on ASW – for both detection and counter-detection operations.[46] Drone swarms deployed at chokepoints (or gateways) of an adversary's docking exit routes could, for example, act as a layered physical barrier, to deter or deny an adversary the ability to operate its submarine within certain military zones.[47] UUVs could also be deployed in real-time for acoustic jamming operations in conjunction with other emitter underwater platforms akin to airborne electronic warfare operations against radar systems.[48] Moreover, UUVs could be used as decoys to create false targets to support counter-detection efforts.

From a tactical perspective, drone swarms would not need ocean-wide coverage (or 'full ocean transparency') to detect and track submarines effectively. According to UK Rear Admiral John Gower, a relatively even spread of sensors might be sufficient to enable "a *viable search and detection plan* could be conceived for the open ocean" (emphasis added).[49] He suggests that a relatively even spread of sensors might be sufficient to enable "a viable search and detection plan … conceived for the open ocean."[50] Furthermore, advances in mobile sensing platforms could enable drones in swarms to locate submarines through chokepoints as they emerge from ports. For example, China's relatively small fleet of SSBNs already faces challenges in leaving port and avoiding key chokepoints, such as the Luzon Strait or the seas surrounding the Ryukyu Islands, without being detected by the US Navy as they transit into the waters of the Indo-Pacific.

Because of the slowness of UUVs today, together with the problem of battery power in extended geographical ranges and timeframes, trailing them – or leech-like UUVs that could attach itself to a submarine – autonomously, for now, seems implausible.[51] Future iterations of ML-augmented UUVs and USVs may eventually complement, and perhaps replace entirely, the traditional role of general-purpose nuclear-powered submarines and manned surface vehicles, in tracking and trailing submarines of adversaries at chokepoints while mounting sparsely distributed and mobile distributed network systems sensors on UUVs.[52]

If a state views the credibility of its survivable nuclear weapons (especially nuclear-armed submarines) to be at risk, conventional capabilities such as drone swarms will likely have a destabilizing effect at a strategic level.[53] Therefore, a drone swarm operation designed to track and monitor SSBNs in ISR only (i.e. defensive) missions, the destabilizing effects on deterrence would likely be similar to a weaponized drone attack.[54] In short, even if swarm sorties were not intended as – or indeed technically capable of – a disarming first strike, the *perception alone* of the feasibility of such an operation would likely elicit distrust between nuclear-armed adversaries and be destabilizing, nonetheless, in particular where asymmetries exist between adversaries.[55]

Besides, the speed of AI could put the defender at a distinct disadvantage, creating additional incentives to strike first (or pre-emptively). In an asymmetric relationship, the vulnerabilities created by technologies such as drone swarms make a pre-emptive strike a very enticing option for a less capable nation, who may view its inferior capabilities as vulnerable to an adversary's exploitation.[56] Moreover, a wider capabilities asymmetric gap between adversaries can cause destabilizing dynamics during peacetime and, potentially, escalatory pressures during a crisis.[57] In sum, the deployment of new military technology in the nuclear domain affects states differently: it will depend on the relative strength of their strategic force structure.[58]

For instance, conceptually speaking, US AI-enhanced UUVs could threaten China's nuclear ballistic and non-nuclear attack submarines.[59] Thus, even if US UUVs were programmed only to threaten China's non-nuclear (or non-strategic) attack submarine fleets, Chinese commanders might nonetheless fear that their country's nascent, and relatively noisy and small – compared to US and Russian SSBN fleets – sea-based nuclear deterrent could be neutralized more easily.[60] Absent US empathy to de-escalate tensions (i.e. "putting oneself in the other fellow's place"),[61] and advances in ML sensor technology – designed to enable more accurate detection of Chinese SSBNs – will likely reinforce Beijing's fear that a superior nuclear-armed power is deliberately targeting it, thus intensifying security dilemma dynamics.[62] To be sure, China's pursuit of capabilities such as AI, quantum computing, and sophisticated C3I systems – which US commanders have railed against – are precisely designed to restore Chinese confidence in their ability to ensure reliable communications with their SSBNs, without being compromised by US anti-submarine (especially UUVs) capabilities.[63]

Enhanced by sophisticated ML neural networks, Chinese manned and unmanned drone teaming operations could potentially impede future US freedom of navigation operations in the South China Seas.[64] Were China to infuse its cruise missiles and hypersonic glide capabilities with AI and autonomy, close-range encounters in the Taiwan Straits and the East and South China Seas would become more complicated, accident-prone, and destabilizing – at both a conventional and nuclear level.[65] Conceptually, US AI-enhanced UUVs could threaten *both* China's nuclear ballistic and non-nuclear attack submarines. In response to this perceived risk, the Chinese navy is developing and deploying UUVs to bolster its underwater monitoring and anti-submarine capabilities, as part of a broader goal to establish an 'underwater Great Wall' to challenge US undersea military primacy.[66]

## Conclusion

This chapter analyzed how AI-augmented and enhanced intelligence gathering and analysis systems might impact the survivability and credibility of states' nuclear deterrence capabilities. The first section found that because of the inherent difficulty in finding mobile missiles, even modest improvements towards enabling this capability – or even the perception that advances of this kind are imminent – could be a strategic game-changer. Thus, irrespective of whether future breakthroughs in AI produce irrefutable evidence of a game-changing means to locate, target, and destroy mobile missile forces, Chinese perception of US intentions in the pursuit of these capabilities would, therefore, be far more crucial.

For example, if AI-enabled improvements in the accuracy and remote sensing of counterforce capabilities enabled the US to surmount China's deception and dispersal tactics, and locate its mobile nuclear missiles, the implications for China–US crisis stability would be significant. As described in chapter 4, despite China's no-first-use policy, Beijing might feel impelled to engage in limited nuclear escalation to reclaim escalation dominance. Moreover, because of the tendency of Chinese strategists to extrapolate from current US capabilities, malign intent (and to assume future ones will threaten their security), even incremental improvements in AI techniques to integrate and synthesize data about the location of an adversary's mobile missiles would exacerbate pre-existing fears and distrust.[67]

This analysis revealed that the capabilities AI might *enhance*, together with the ones it could *enable*, might make hunting for mobile nuclear arsenals faster, cheaper, and more effective than before. Thus, AI-enabled and enhanced capabilities will likely have a more significant impact on strategic stability than the sum of its parts. An increase in the vulnerability of states' nuclear mobile systems caused by AI innovations might upend the deterrence assumptions that undergirded the nuclear revolution – relying on hardening and concealing approaches for ensuring the survivability and credibility of their nuclear forces.

The second section found that recent advances in sensor, communication, and processing technologies (i.e. big-data analytics and ML) could have a transformative impact on future iterations of ASW and undersea support platforms, thereby significantly enhancing their ability to locate, track, and destroy submarines in real-time such as detecting and cueing torpedo seekers in long-range ASW in 'fire and forget' mission. For now, the technical feasibility of this hypothesis remains highly contested.

While some experts argue that emerging technologies like AI, quantum communications, and big-data analytics may empower new iterations of highly portable sensing, communications, and signal-processing technology, others remind alarmists of the danger of assuming that disruptive technologies will *necessarily* have transformative strategic effects. For one thing, significant advances in power, sensor technology, and communications would be needed *before* autonomous systems have a game-changing strategic impact on submarine reconnaissance and deterrence.

Ultimately, AI's impact will depend on the relative strength of states' strategic force structure. A wider capabilities asymmetric gap between adversaries can cause destabilizing dynamics during peacetime and escalatory pressures during a crisis. Advances in ML sensor technology for enabling more accurate detection of Chinese SSBNs might, for example, reinforce Beijing's fear that it was being targeted by a militarily superior power (i.e. the US).

In sum, future iterations of AI that can make predictions using expanded and dispersed data-sets, and then locate, track, and target states' strategic missiles forces in underground silos – and in particular, mobile ICBM launchers – onboard stealth aircraft, SSBNs, and truck or rail-mounted, is set to grow.[68] How and why might AI-augmented autonomous weapon systems such as drone swarms and hypersonic vehicles impact nuclear stability? It is to this issue that we turn to in chapter 6.

## Notes

1 James S. Johnson, "Artificial Intelligence: A Threat to Strategic Stability," *Strategic Studies Quarterly*, 14, 1 (2020), pp. 16–39.
2 Technological advancements associated with the 'computer revolution' (e.g. guidance systems, sensing technology, data processing, quantum communication, big-data analytics, and AI) have already improved the robustness of counterforce capabilities and, in turn, reduced the survivability of mobile missiles and submarines. Keir A. Lieber and Daryl G. Press, "Why States Won't Give Nuclear Weapons to Terrorists," *International Security*, 38, 1 (2013), pp. 80–104.
3 Alfred T. Mahan, *Armaments and Arbitration: Or, The Place of Force in the International Relations of States* (New York: Harper & Brothers, 1912), p. 105.
4 Keir A. Lieber and Daryl G. Press, "The New Era of Counterforce: Technological Change and the Future of Nuclear Deterrence," *International Security* 41, 4 (2017), pp. 9–49.
5 'Autonomy' is fundamentally a software endeavor. That is, software (i.e. AI ML techniques for sensing, modeling, and decision-making) rather than hardware

separates existing armed unmanned and remote-controlled weapon systems (e.g. the US MQ-9 Reaper) from 'fully autonomous' iterations.
6 For example, in 2019, an Israeli defense company Rafael Advanced Defense Systems announced the development of a new ML AI-augmented ATR capability, for use in the final stages of guidance to track a pre-determined target. "Rafael Unveils New Artificial Intelligence and Deep-Learning Technologies in SPICE-250 to Enable Automatic Target Recognition," *Rafael Advanced Defense Systems Ltd.*, June 10, 2019, www.rafael.co.il/press/elementor-4174/ (accessed June 11, 2019).
7 In addition to AI, vulnerabilities to mobile missiles systems may also include: (1) exaggerating how 'mobile' the missile systems are; (2) geographic constraints of mobile systems (i.e. land, sea, air); (3) mobile command and control systems can be hacked; (4) outsized alert signatures; high demands on crews; operational security is hard to secure (i.e. information leaks or 'tells'); and (5) insider attacks on mobile missiles are a constant danger. Paul Bracken, "The Cyber Threat to Nuclear Stability," *Orbis* 60, 2 (2016), pp. 188–203, p. 194.
8 This analysis focuses on land-based mobile missiles deployed on transporter-erector-launchers (TELs) or railroads. TELs are tracked or wheeled vehicles that move on or off-road and can be quickly prepped to launch ballistic missiles.
9 Examples of intelligence collection systems that may benefit from big-data analysis and AI algorithmic decision support include Communications Intelligence, Electronic Intelligence, and Imagery Intelligence.
10 Analysts continue to emphasize the various technical challenges in locating mobile missiles for counterforce operations. UAVs would need to track in real-time and communicate in real-time in a manner that could not be detected and cue a rapid surprise attack *before* the mobile launchers can be relocated. Austin Long and Brendan Rittenhouse Green, "Stalking the Secure Second Strike: Intelligence, Counterforce, and Nuclear Strategy," *Journal of Strategic Studies*, 38, 1–2 (2015), pp. 21–24.
11 DARPA, "ACTUV Sea Hunter prototype transitions to the US Office of Naval Research for further development," January 30, 2018, www.darpa.mil/news-events/2018-01-30a (accessed June 10, 2019).
12 Jonathan Gates, "Is the SSBN Deterrent Vulnerable to Autonomous Drones?" *The RUSI Journal*, 161, 6 (2016), pp. 28–35.
13 While there are several technologies under development specifically designed to track SSBNs (e.g. the DoD's *Sea Hunter*, a prototype autonomous surface vehicle), these programs are immature. Several technical challenges remain in the development of ASW, limiting their operational utility as weapons for offensive operations over extended geographical ranges and duration – above all, battery power. Supplying sufficient power for swarms of UAVs (or UUVs) for an extended period would require significant improvements in either battery technology, air-independent propulsion, or fuel-cell technology. Moreover, many states' nuclear-related facilities (except the SSBNs) are located well inland, which (for now) makes drones ill-suited to attack these targets, unless lifted in by a different platform. However, electric storage battery power capacity is

rapidly improving, and experts predict a 10-fold increase in power and endurance within the next decade. See Leslie F. Hauck and John P. Geis II, "Air Mines: Countering the Drone Threat to aIRcraft," *Air & Space Power Journal*, 31, 1 (Spring 2017), pp. 26–40; and Gates, "Is the SSBN Deterrent Vulnerable to Autonomous Drones?" pp. 28–35; and Sebastian Brixey-Williams, "Will the Atlantic Become Transparent?" 2nd ed. *British Pugwash*, November 2016, https://britishpugwash.org/wp/wp-content/uploads/2016/11/Will-the-Atlantic-become-transparent-.pdf (accessed March 10, 2020).

14 A nuclear-armed state would need a very high degree of confidence that it could identify and pre-emptively destroy (or disable) all of an adversary's nuclear-weapon delivery systems capable of launching devastating retaliatory attacks. If a counterforce attack intends to disarm the adversary *before* it can respond with nuclear weapons, targeting certainty would need to be almost 100%. Joseph Johnson, "MAD in an AI Future?" Center for Global Security Research (Livermore, CA: Lawrence Livermore National Laboratory), pp. 4–6.

15 Li Bin, "Tracking Chinese Strategic Mobile Missiles," *Science and Global Security*, 15 (2007), pp. 1–30; Charles L. Glaser and Steve Fetter, "Should the United States Reject MAD? Damage Limitation and US Nuclear Strategy toward China," pp. 49–98.

16 Avery Goldstein, "First Things First: The Pressing Danger of Crisis Instability in US–China Relations," *International Security*, 37, 4 (Spring 2013), pp. 49–89.

17 Caitlin Talmadge, "Would China Go Nuclear? Assessing the Risk of Chinese Nuclear Escalation in a Conventional War with the United States," *International Security*, 41, 4 (Spring 2017), pp. 90–91.

18 Tong Zhao and Li Bin, "The underappreciated risks of entanglement: a Chinese perspective," in James M. Acton (ed.), with Li Bin, Alexey Arbatov, Petr Topychkanov, and Zhao Tong, *Entanglement: Russian and Chinese Perspectives on Non-Nuclear Weapons and Nuclear Risks* (Washington, DC: Carnegie Endowment for International Peace, 2017), pp. 47–75.

19 Talmadge, "Would China Go Nuclear?" pp. 50–92.

20 Experts have long stressed the importance of separating nuclear detection and early warning systems from other parts of the nuclear command and control chain to prevent system accidents from these interactions. See Scott D. Sagan, *The Limits of Safety: Organizations, Accidents, and Nuclear Weapons* (Princeton, NJ: Princeton University Press, 1993).

21 Goldstein, "First Things First: The Pressing Danger of Crisis Instability in US–China Relations," pp. 67–68.

22 Jon R. Lindsay and Erik Gartzke (eds), *Cross-Domain Deterrence: Strategy in an Era of Complexity* (Oxford: Oxford University Press, 2019), p. 19.

23 Crisis instability and misinterpreted warnings are largely determined by the perceptions (or rather misperceptions) of an adversary's intentions prompting an incidental strike or threat. See James M. Acton, "Escalation Through Entanglement: How the Vulnerability of Command-and-Control Systems Raises the Risks of an Inadvertent Nuclear War," *International Security*, 43, 1 (2018), pp. 56–99, p. 93.

24 Robert Jervis, *The Meaning of the Nuclear Revolution: Statecraft and the Prospect of Armageddon* (Ithaca, NY: Cornell University Press, 1989).
25 Paul Scharre, "Autonomous Weapons, and Operational Risk – Ethical Autonomy Project" (Washington, DC: Center for a New American Security, November 2017), p. 33.
26 For example, in 2018, Russian President Putin stated that Russia's AI-enhanced weapons are "invincible against all [i.e. US] existing and prospective missile defense and counter-air defense systems." August Cole and Amir Husain, "Putin Says Russia's New Weapons Can't Be Beat. With AI and Robotics, They Can," *Defense One*, March 13, 2018, www.defenseone.com/ideas/2018/03/putin-says-russias-new-weapons-cant-be-beat-ai-and-robotics-they-can/146631/ (accessed June 10, 2019).
27 Barry R. Posen, *Inadvertent Escalation: Conventional War and Nuclear Risks* (Ithaca, NY: Cornell University Press, 1991), chapter 1.
28 The nuclear threshold could also be breached if an adversary felt compelled to use its weapons before being disarmed in retaliation for an unsuccessful disarming strike, or if a conflict or crisis triggered accidental deployment.
29 Eric Heginbotham et al., *The US–China Military Scorecard: Forces, Geography, and the Evolving Balance of Power, 1996–2017* (Santa Monica, CA: RAND Corporation, 2015), p. 353.
30 Successfully targeting mobile missile TELs while they are moving is technically very difficult. To launch its missile, the TEL must stop for a brief period, during which it is most vulnerable. Because this vulnerability window is often short, attacking weapons either have to be very close to their target or travel at very high speeds. As chapter 7 will show, militaries are developing hypersonic technology to address this very issue.
31 Offensive ASW tactics which used long-range sensors in wide-area submarine detection and tracking operations are often referred to as 'making the ocean transparent.'
32 The US DARPA is currently developing an anti-submarine warfare continuous trail unmanned vehicle capability, the Anti-Submarine Warfare Continuous Trail Unmanned Vessel program, to track quiet diesel-electric submarines with USVs from the surface. In 2017, China reportedly launched a new stealthy unmanned oceanic combat vehicle (the D3000), capable of engaging in both anti-submarine and surface warfare missions. P.W. Singer and Jeffrey Lin, "With the D3000, China enters the robotic warship arms race," *Popular Science*, September 25, 2017, www.popsci.com/robotic-warship-arms-china-d3000/ (accessed June 10, 2019).
33 Less benign environments, such as trying to access Arctic Ice or contested anti-access and area denial zones, would be far more complicated and, during crisis and conflict, potentially escalatory and accident-prone. A submarine commander could reduce a submarine's vulnerability to ASW operations in the following ways: using the thermocline, changing speed, depth, heading, bathymetric features (i.e. the depth of water relative to sea level), and the hunting ship's acoustic profile, as well as using decoys and surface or environmental

disturbance techniques. David Blagden, "What DARPA's Naval Drone Could Mean for the Balance of Power," *War on the Rocks*, July 9, 2016, https://warontherocks.com/2015/07/what-darpas-naval-drone-could-mean-for-the-balance-of-power/ (accessed March 10, 2020).

34 Unmanned drone platforms are capable of carrying several types of sensors, and the swarming ML systems to control them have either available today, or in advanced stages of development, active and passive sonar, magnetic anomaly detectors, light detection and ranging (LIDAR) systems, thermal sensors, and laser-based optical sensors capable of piercing seawater.

35 For example, in 2018, a team at Newcastle University in the UK developed ultra-low-cost acoustic 'nano models,' which can send data via sound up to two kilometers in short-range underwater networks. See "A Better Way to Transmit Messages Underwater," *The Economist*, May 12, 2018, www.economist.com/science-and-technology/2018/05/12/a-better-way-to-transmit-messages-under water (accessed June 10, 2019).

36 Even failed ASW operations have compelled a submarine to evade and lose the initiative, or made it more traceable, for a fresh ASW attack. Bryan Clark, *The Emerging Era in Undersea Warfare* (Washington, DC: Center for Strategic and Budgetary Assessments, 2018), pp. 3–4.

37 Owen R. Cote Jr., "Invisible Nuclear-Armed Submarines, or Transparent Oceans? Are Ballistic Missile Submarines Still the Best Deterrent for the United States?" *Bulletin of the Atomic Scientists*, 75, 1 (2019), pp. 30–35, p. 30.

38 Ibid., p. 10.

39 Brixey-Williams, "Will the Atlantic Become Transparent?"

40 Today, only the US possesses the requisite anti-submarine warfare capabilities and global maritime scale to render quiet submarines vulnerable. See Cote, "Invisible Nuclear-Armed Submarines, or Transparent Oceans?" p. 33.

41 Finally, unlike standard UUVs, which are typically tethered and have short ranges, underwater gliders (e.g. US Liquid Robotics *Waverider* SV3), while slow, can roam over long distances for months at a time. Gates, "Is the SSBN Deterrent Vulnerable to Autonomous Drones?" pp. 28–35; and Bradley Martin, Danielle C. Tarraf, Thomas C. Whitmore, Jacob DeWeese, Cedric Kenney, Jon Schmid, and Paul DeLuca, *Advancing Autonomous Systems: An Analysis of Current and Future Technology for Unmanned Maritime Vehicles* (Santa Monica, CA: RAND Corporation, 2019).

42 Given the vast geographical distances involved, new sensing and signaling technologies would need to be developed and deployed without the aid of local ground-based centers or airborne sensor platforms for gathering and processing data. Gates, "Is the SSBN Deterrent Vulnerable to Autonomous Drones?" pp. 28–35.

43 See Aleem Datoo and Paul Ingram, "A Primer on Trident's Cyber Vulnerabilities," *BASIC* Parliamentary Briefings on Trident Renewal Briefing 2 (March 2016), www.basicint.org/wp-content/uploads/2018/06/BASIC_cyber_vuln_mar2016.pdf (accessed June 10, 2019).

44 For example, SSBNs did not become a universally survivable nuclear asset, as

passive acoustics did not render the entire fleets of SSBNs vulnerable. Cote, "Invisible Nuclear-Armed Submarines, or Transparent Oceans?" p. 33.

45 Experts anticipate technical advances in battery and fuel cell technology will shortly enable non-nuclear submarines, UUVs, and other undersea systems to conduct long-duration ASW operations in extended ranges. For example, see Alan Burke, "System modeling of an Air-Independent Solid Oxide Fuel Cell System for Unmanned Undersea Vehicles," *Journal of Power Sources*, 158, 1 (July 2006), pp. 428–435.

46 In 2018, the US Office of Naval Research requested a white paper to study analytical research on the relationship between physical oceanographic changes and sound transmission, including the development and use of AI and ML techniques to collect relevant data-sets. Patrick Tucker, "How AI Will Transform Anti-Submarine Warfare," *Defense One*, July 1, 2019, www.defenseone.com/technology/2019/07/how-ai-will-transform-anti-submarine-warfare/158121/ (accessed June 10, 2019).

47 Most modern submarines (notably louder types) would be unable to pass through confined chokepoints to get within range of their missiles' targets undetected by active sonar-based techniques. Ibid., p. 35.

48 Ning Han, Xiaojun Qiu, and Shengzhen Feng, "Active Control of Three-Dimension Impulsive Scattered Radiation Based on a Prediction Method," *Mechanical Systems and Signal Processing* (July 30, 2012), pp. 267–273.

49 John Gower, "Concerning SSBN Vulnerability – Recent Papers," *BASIC*, June 10, 2016, www.basicint.org/blogs/rear-admiral-john-gower-cbobe/06/2016/concerning-ssbn-vulnerability-%C2%AD-recent-papers (accessed June 10, 2019).

50 The leech-like UUV could carry a transponder that may increase the sonar signature of the target submarine and help identify it. See Norman Friedman, "Strategic Submarines and Strategic Stability: Looking Towards the 2030s," *National Security College, Crawford School of Public Policy ANU College of Asia & the Pacific*, September 2019, https://nsc.crawford.anu.edu.au/publication/15176/strategic-submarines-and-strategic-stability-looking-towards-2030s (accessed June 10, 2019).

51 It might be possible for a hand-off to occur between drones in a grid to monitor a submarine as it moves, but doing so in extended ranges and duration would be cumbersome and slow.

52 To date, the US Navy has deployed and tested digital network systems in littoral waters. For example, PLUSNet (Persistent Littoral Undersea Surveillance Network) is a joint project between the US Navy's Office of Naval Research and DARPA that began in 2005. 'Persistent Littoral Surveillance: Automated Coast Guards,' *Naval Technology*, April 30, 2012, www.naval-technology.com/features/featurenavy-persistent-littoral-surveillance-auvs-uuvs/ (accessed July 10, 2019).

53 An asymmetric encounter involving adversaries who do not possess ASW capabilities, the escalatory cycles described above would unlikely occur. Ibid., p. 132.

54 Jurgen Altmann and Frank Sauer, "Autonomous Weapons and Strategic Stability," *Survival*, 59, 5, (2017), pp. 121–127, p. 131.
55 Arguably, by improving UAVs' ability to identify SSBNs more accurately than before, AI technology may reduce the risks of accidental collisions and other accidents – especially involving drones – in at-sea deterrence missions.
56 See Steven Metz and James Kievit, *Strategy and the Revolution in Military Affairs: From Theory to Policy* (Carlisle: Strategic Studies Institute 1995).
57 See Bryan R. Early and Erik Gartzke, "Spying from Space: Reconnaissance Satellites and Interstate Disputes," *Journal of Conflict Resolution* (March 2021). https://doi.org/10.1177/0022002721995894.
58 For example, because China has yet to develop a long-range bomber force and relies on ballistic missiles for its strategic deterrent, US ASW capabilities would likely be viewed as threatening to China.
59 A range of autonomous ground vehicles and underwater vehicles are already in development globally with varying degrees of success. Mary L. Cummings, *Artificial Intelligence and the Future of Warfare* (London: Chatham House, 2017), pp. 8–9.
60 For example, Chinese reports from the 2016 seizure of a US UUV suggest that this action was taken because of the perceived threat posed to Chinese SSBNs by the US Navy in the region.
61 John H. Herz, *International politics In the Atomic Age* (New York: Columbia University Press, 1959), p. 249.
62 From open sources, limited information is available about China's SSBN C2 protocols or communications systems. Western analysts expect that if the PLA Navy keeps its SSBNs on continuous at-sea deterrent patrols in the future, it will likely conduct those patrols close to Chinese shores and protect its submarines using conventional naval capabilities. Once China develops a next-generation SSBN that is quieter than its current-generation Type-094 boat, it may shift to an open ocean mode of deployment. Wu Riqiang, "Have China's Strategic Nuclear Submarines Already Commenced Operational Patrols?" *Dangdai Jianchuan* [Modern Ships], 1 (2016), p. 34.
63 For example, the Commander of US Strategic Command, General John Hyten, has expressed his concerns publicly about China's pursuit of quantum computing and communications military capabilities. General John E. Hyten, Statement before the House Committee on Armed Services, Washington DC, March 28, 2019, www.armed-services.senate.gov/imo/media/doc/Hyten_02-26-19.pdf (accessed July 10, 2019).
64 China's China Aerospace Science and Industry Corporation is currently developing a stealth drone (*Skyhawk*) that will reportedly be capable of sharing data directly with manned aircraft. Kristin Huang, "China's Sky Hawk stealth drone can 'talk' to fighter pilots, the developer says," *South China Morning Post*, January 11, 2019, www.scmp.com/ news/china/military/article/2181731/chinas-sky-hawkstealth-drone-has-capability-talk-fighter-pilots (accessed July 10, 2019).
65 The South China Sea is considered an attractive bastion for strategic submarines because its hydrographic conditions (i.e. depth, temperature, and salinity) are

considered particularly difficult for ASW operations. Recent reports indicate that China is engaged in the development of several potentially destabilizing capabilities, including research into the use of AI and autonomy in prompt and high-precision (cruise and ballistic) missile systems, space planes, and a variety of hypersonic guide vehicles. Office of the Secretary of Defense, *Annual Report to Congress: Military and Security Developments Involving the People's Republic of China, 2019* (Washington, DC: US Department of Defense, 2019), https://media.defense.gov/2019/May/02/2002127082/-1/-1/1/2019_CHINA_MILITARY_POWER_REPORT.pdf (accessed July 10, 2019).

66 Today, the main threat posed to the US from China is to US attack submarines operating in forward ranges in Chinese coastal waters, not to its SSBNs operating at greater ranges in the Pacific. Clark, *The Emerging Era in Undersea Warfare*.

67 James S. Johnson, "Washington's Perceptions and Misperceptions of Beijing's Anti-Access Area-Denial 'Strategy': Implications for Military Escalation Control and Strategic Stability," *The Pacific Review*, 20, 3 (2017), pp. 271–288.

68 In addition to UAVs, emerging space technologies will soon enable drone-like surveillance from space, incorporating similar ML techniques. Larger satellite constellations, coupled with smaller individual satellites, is expected to provide continuous coverage over large geographical ranges. "How AI could destabilize nuclear deterrence," Elias Groll, April 24, 2018, https://foreignpolicy.com/2018/04/24/how-ai-could-destabilize-nuclear-deterrence/ (accessed March 10, 2020).

# 6

# The fast and the furious: drone swarming and hypersonic weapons

How might AI-augmented drone swarming and hypersonic weapons complicate missile defense, undermine states' nuclear-deterrent forces, and increase the risk of escalation?[1] How might AI-augmented unmanned systems effect escalation, deterrence, and conflict management, when fewer human lives are perceived to be at risk? The proliferation of a broad range of AI-augmented autonomous weapon systems – most notably drones used in swarming tactics – might have significant strategic implications for nuclear security and escalation in future warfare.[2] Unmanned autonomous systems (UAS) could be deployed in complex missions in hitherto inaccessible and cluttered environments (e.g. under-sea anti-submarine warfare), and aerial and underwater drones in swarms might eventually replace intercontinental ballistic missiles (ICBMs) and nuclear-powered ballistic missile submarines (SSBNs) for the delivery of nuclear weapons.[3]

Several observers anticipate that sophisticated AI-augmented UAS (both weaponized and unarmed) will soon be deployed for a range of intelligence, surveillance, and reconnaissance (ISR) and strike missions.[4] Even if UAS are used only for conventional operations, their proliferation could have destabilizing implications and increase the risk of inadvertent nuclear escalation. For example, AI-augmented drone swarms used in offensive operations targeting ground-based air defenses, used by nuclear-armed states to defend their strategic assets (e.g. launch facilities, and early warning, and attendant nuclear command, control, and communications systems), might pressure a weaker nuclear power to respond with nuclear weapons in a use-them-or-lose-them situation.[5]

As demonstrated in Part II, these advances have significantly increased the perceived operational value great military powers attach to the development of a range of UAS (i.e. ground-based, air-borne, on-sea, and under-sea drones), thus making the delegation of lethal authority to UAS an increasingly irresistible and destabilizing prospect.[6] That is, to defend or capture the technological upper hand in possession of cutting edge warfighting assets vis-à-vis strategic rivals, traditionally conservative militaries may eschew the potential risks of deploying unreliable, unverified, and unsafe UAS.[7] In short, immature deployment of these nascent systems in a nuclear context could have severe consequences.[8]

The remainder of this chapter proceeds as follows. First, it describes how AI augments existing iterations of UAS. Next, it considers the trade-off states face between increasing speed and precision, and the uncertainties and risks of deploying potentially accident-prone, unpredictable, and unreliable UAS without adequate human control and responsibility. Second, the chapter unpacks the possible strategic operations (both offensive and defensive) that AI-augmented drone swarms might execute, and the potential impact of these operations for crisis stability. Finally, it examines how machine learning (ML)-enabled qualitative improvements to hypersonic delivery systems (i.e. hypersonic guide vehicles (HGVs), hypersonic scramjets, and hypersonic cruise missiles) might amplify the escalatory effects of long-range – conventional and nuclear-armed – precision munitions.

## AI-enhanced drone swarms: the danger of taking humans out of the loop

Conceptually speaking, autonomous systems will incorporate AI technologies such as visual perception, speech, facial and image recognition, and decision-making tools to execute a range of core air interdiction, amphibious ground assaults, long-range strike, and maritime operations independent of human intervention and supervision.[9] Currently, only a few weapon systems select and engage their targets without human intervention. Loitering attack munitions (LAMs) – also known as 'loitering munitions' or 'suicide drones' – pursue targets (such as enemy radars, ships, or tanks) based on preprogrammed targeting criteria, and launch an attack when their sensors detect an enemy's air defense radar.[10] Compared to cruise missiles (designed to fulfill a similar function), LAMs use AI technology to shoot down incoming projectiles faster than a human operator ever could and can remain in flight (or loiter) for much longer periods. This attribute could complicate the ability of states to reliably and accurately detect and attribute autonomous attacks.[11]

A low-cost lone wolf drone, for example, would unlikely pose a significant threat to a US F-35 stealth fighter.[12] However, hundreds of AI ML autonomous drones in a swarming sortie might evade and overwhelm an adversary's sophisticated defense capabilities – even in heavily defended regions such as China's east and coastal regions. Further, stealth variants of these systems (armed with miniaturized electromagnetic jammers and cyber-weapons), could be used to interfere with or subvert an adversary's targeting sensors and communications systems,[13] thereby undermining its air-defenses in preparation for an offensive strike.[14]

In 2011, for example, US Creech Air Force Base aircraft cockpit systems – operating MQ-1 and MQ-9 unmanned drones in the Middle East – were infected with hard-to-remove malicious malware, exposing the vulnerability of US cyber-attacks.[15] However, this threat may be countered by integrating future iterations of AI into manned stealth fighters such as the F-35.[16] Manned F-35 fighters will soon be able to leverage AI-powered robotics to control small drone swarms close to the aircraft to execute sensing, reconnaissance, and targeting operations – such as countermeasures against swarm attacks. In the future, pilots will be capable of operating their teamed drones directly from the cockpit and, eventually, command teams of 'wingman drones' simultaneously.[17] Alternative countermeasures against drone swarms that will likely parallel development in swarm technology include radars to detect and track swarms, and high-energy lasers to destroy them.[18] Extended endurance of unmanned aerial vehicles (UAVs) and support platforms could potentially increase the ability of drone swarms to survive these kinds of countermeasures, however.

According to former US Deputy Secretary of Defense Robert Work, the US, "will not delegate lethal authority for a machine to make a decision" in the use of military force. Work added, however, that such self-restraint could be tested if there was a strategic near-peer competitor (notably China and Russia) "who is more willing to *delegate authority* to machines than we are and, as that competition unfolds, we'll have to make decisions on how we can best compete" (emphasis added).[19] As chapter 8 will show, the pre-delegation of authority to machines, and taking human judgment further out of the crisis decision-making process, might severely challenge the safety, resilience, and credibility of nuclear weapons in future warfare.[20]

The historical record is replete with examples of near nuclear misses, demonstrating the importance of human judgment in mitigating the risk of miscalculation and misperception (i.e. of another's intentions, red-lines, and willingness to use force) between adversaries during crises.[21] Eschewing these risks, China has incorporated a range of advanced UAVs into all four services of its armed forces.[22] Further, China plans to incorporate AI into UAVs and UUVs for swarming missions infused with AI ML technology. At a diplomatic level, Beijing has signaled its support in principle for a ban on lethal autonomous weapons (LAWs) but is simultaneously engaged in actively developing military applications of AI and autonomy.[23] For example, China has researched data-link technologies for 'bee-swarm' UAVs, emphasizing network architecture, navigation, and anti-jamming capabilities for targeting US aircraft carriers.[24]

Military commanders value informational awareness and the psychological control over decisions during crises above all else. Accident-prone UAS should, therefore, persuade militaries to counsel caution in the use of these capabilities. The intense psychological preference for control and organizational politics will not only influence how militaries perceive the utility of UAS during wartime but also decisions on deterrence and arms control (or arms racing) during peacetime.[25] As described in chapter 2, escalation risks associated with fear and risk aversion can stabilize or destabilize outcomes. A crucial distinction is, therefore, between the risk of *unintentional* escalation and *intentional* escalation.

Today, the risks associated with unpredictable AI-augmented autonomous systems operating in dynamic and complex – and possibly a priori unknown environments – is underappreciated by global defense communities.[26] International dialogue has improved mutual understanding of LAWS – e.g. unmanned combat aerial vehicles, smart ammunition systems, and combat robots – and achieved general agreement on the importance of human control in the use of force at a conceptual level. However, complicated definitional issues over what exactly constitutes a LAWS, and how any policies might be enforced, supervised, controlled, and verified have yet to be resolved. Moreover, despite these efforts, there seems to be no shared understanding of what these concepts mean in practice or how they would be operationalized.[27] Open questions related to these issues include: what is an appropriate level of interaction between humans and autonomous weapon systems, and for how long, and under what circumstances, should humans trust a machine to operate without human intervention?

The current discourse surrounding LAWs does not adequately consider the potentially significant implications of the use of *unarmed* autonomous drones capable of sophisticated ISR operations in adversarial territories.[28] Advances in autonomy and AI described in Part I make the strategic challenges described below all the more pressing. In sum, the uncertainties and unpredictability surrounding the deployment of accident-prone UAS operating at machine speed, and competitive geopolitical pressures impelling the increasing delegation of control to opaque (and dual-use) algorithms, will likely increase crisis stability and exacerbate security dilemmas.[29]

## Swarming and new strategic challenges

Drones used in swarms are *conceptually* well suited to conduct pre-emptive attacks and nuclear-ISR missions against an adversary's nuclear mobile missile launchers and SSBNs, and their enabling facilities (e.g. command, control, communications, and intelligence (C3I) and early warning systems, antennas, sensors, and air intakes).[30] For example, the Defense Advanced Research Projects Agency (DARPA)'s autonomous surface vehicle prototype, *Sea Hunter*, is designed to support anti-submarine warfare operations, including submarine reconnaissance.[31] As chapter 5 described, some observers fear that autonomous systems like the *Sea Hunter* might render the underwater domain 'transparent,' thus eroding stealthy SSBNs' deterrence utility. In short, the ability of future iterations of AI to make predictions based on the synthesis of expanded and dispersed data-sets to locate, track, and target nuclear weapons (e.g. ICBM launchers in underground silos and onboard stealth bombers and SSBNs) is set to grow.[32]

Irrespective of their current technical feasibility, autonomous systems like DARPA's *Sea Hunter* demonstrate how the emerging generation of autonomous weapons is expediting the completion of the iterative targeting cycle to support joint operations, thus increasing the uncertainty about the reliability and survivability of states' nuclear second-strike capability, and potentially triggering use-them-or-lose-them situations.[33] Conceptually, the most destabilizing impact of AI on nuclear deterrence would be the synthesis of autonomy with a range of ML-augmented sensors, undermining states' confidence in the survival of their second-strike capabilities and in extremis triggering a retaliatory first strike. Further, the interplay of the risk factors described above can contribute to the risk of escalation, either by incentivizing first-mover behavior or by exacerbating the risk of misperception and miscalculation between nuclear-armed rivals during a crisis or conventional conflict.[34]

To illustrate these dynamics: State A uses a drone over an adversary's territory on a clandestine ISR mission, which State B perceives as a violation of its sovereignty, or a precursor for an attack planned by A. In this way, the use of UAVs between rivals can introduce pre-emptive risks and create first-mover advantage incentives. For instance, the discovery of information by State A (e.g. suspicious troop movement or missiles leaving their garrison) could incentivize it to escalate a situation, motivated by the desire to capture the strategic upper hand (or terminate a conflict) *before* B can respond. Moreover, State A might consider the first-mover advantages of using low-cost and relatively dispensable drones as controllable at a conventional level, which at least in the opening stages of a conflict would likely increase their appeal. In the event State A's UAV is shot down by B, however, A would be forced to either accept the loss of this vulnerable asset or escalate a situation.[35]

In sum, enhanced by the exponential growth in computing performance and coupled with advances in ML techniques (especially ML-augmented remote sensors) that can rapidly process data in real-time, AI will empower drone swarms to perform increasingly complex missions, such as hunting hitherto hidden nuclear deterrence forces (see chapter 5).[36]

The following four scenarios illustrate the possible strategic operations that AI-augmented drone swarms could execute.[37] First, drone swarms could be deployed to conduct ISR operations to locate and track dispersed (nuclear and conventional) mobile missile launchers, and their attendant dual-use C3I systems.[38] Specifically, swarms incorporating AI-infused ISR, autonomous sensor platforms, automatic target recognition (ATR), and data analysis systems may enhance the effectiveness and speed of sensor drones to locate mobile missiles and evade enemy defenses.[39] Then, satellite imagery and signals intelligence from these swarms would cue stealth fighters or armed drones and destroy these missiles. Further, ATR systems may also be used to enable sophisticated collaborative targeting operations by multiple unmanned aircraft, where one drone automatically hands-off targeting information to another armed unmanned drone.[40] In this way, autonomy in advanced AI-augmented drone swarms will likely exacerbate the co-mingling problem-set, thus increasing strategic instability.[41]

Second, swarming may enhance legacy conventional and nuclear weapon delivery systems (e.g. ICBMs and submarine launched ballistic missile (SLBMs)), potentially incorporating hypersonic variants – discussed in the section that follows. It is noteworthy that at least two nuclear-armed states have developed UAV or UUV prototypes with nuclear delivery optionality.[42] AI applications will likely enhance the delivery system targeting and tracking, and improve the survivability of drone swarms against the current generation of missile defenses. Paradoxically, a dependency on swarms in these systems – similar to cyber-defenses examined in chapter 7 – could make them *more vulnerable* to attack (e.g. spoofing, manipulation, digital jamming, and electromagnetic pulses), risking collisions, occlusions, and loss of communication.[43] To reduce these vulnerabilities, sensor drone swarm formations could apply AI-augmented ISR to bolster intelligence collection, and intra-swarm communication and analysis, widening the geographical range of its operations and monitoring potential threats to the swarm, thereby leaving the remainder of the swarm unfettered to perform its offensive activities.[44] For example, DARPA recently tested how drone swarms might collaborate and coordinate tactical decisions in a high-threat environment with minimal (or denied) communications.[45]

Third, swarming tactics could bolster a states' ability to disable or suppress an adversary's defenses – air defenses, missile defenses, and anti-submarine warfare defenses, clearing the path for a disarming attack.[46] Drone swarms might be armed with cyber or electronic warfare capabilities (in addition to anti-ship, anti-radiation, or regular cruise and ballistic missiles) to interfere with or destroy an adversary's early warning detection and C3I systems in advance of a broader offensive campaign.[47] For example, in 2019, a Chinese drone manufacturer, Zhuhai Ziyan, reportedly developed helicopter drones capable of carrying mortar shells, grenades, and machine guns, and operating autonomously in coordinated swarms.[48] Moreover, non-state groups (e.g. ISIS forces in Iraq and Houthis in Yemen) have also experimented with small commercially available multi-rotor drones for directing artillery attacks.[49]

Finally, in the maritime domain, UUVs, unmanned surface vehicles (USVs), and UAVs supported by AI-enabled intra-swarm communication and ISR systems could be deployed simultaneously in *both* offensive and defensive antisubmarine warfare operations to saturate an enemy's defenses and to locate, disable, and destroy its nuclear-armed or non-nuclear attack submarines.[50] Despite continued advances in sensor technology design (e.g. reduced size and extended detection ranges) to overcome quieting challenges, other technical challenges still remain (see chapter 5). These include communicating underwater between multiple systems, processing power requirements, generating battery life and energy, and scaling the system.

In a conventional counterforce operation, a state could, for example, attack an enemy's sensors and control systems with a drone swarm armed with electronic warfare or cyber weapons, degrading its integrated air-defense systems (e.g. for spoofing and electromagnetic pulse attacks).[51] Simultaneously, a state could use a separate drone swarm to draw fire away from its weapon systems and protect its sensors,[52] as a prelude for long-range (manned or unmanned) stealth bombers flanked by conventionally or nuclear-armed drones. Drone swarms, operating in intra-swarm cooperation means of command, might also be used in swarm vs. swarm combat scenarios – including drones armed with dual-payloads and hypersonic variants. Currently, machine–machine collaboration (or interaction) is still at a very nascent stage of development.[53] Conversely, drone swarms might enhance states' missile defenses as countervails to this type of offensive threat. For example, swarms could form a defensive wall to absorb incoming missile salvos, intercepting them or acting as decoys to throw them off course with mounted laser technology.[54] Besides, swarms attacks could be countered by other capabilities, including high-power microwave attacks or large shotguns with small munitions.[55]

Notwithstanding the remaining technical challenges (especially the demand for power), swarms of robotic systems fused with AI ML techniques may presage a powerful interplay of increased range, accuracy, mass, coordination, intelligence, and speed in a future conflict.[56] Perceived as a relatively low-risk *force majeure* with ambiguous rules of engagement – and absent of a robust normative and legal framework – lethal and non-lethal autonomous weapons will likely become an increasingly attractive asymmetric capability to undermine a militarily superior rival's readiness and resolve – both to gain coercive and military advantages and prevent an adversary from deriving benefits from this capability.[57] Autonomous drone systems could, for example, be deployed in 'salami-slicing' tactics to chip away at an adversary's will (or resolve), but without crossing a threshold (or psychological red-line) that would provoke escalation.[58]

## Hypersonic weapons and missile defenses

Multiple advanced non-nuclear weapons could potentially threaten a wide range of strategic targets. In particular, technological advances in hypersonic weapons deployed in conjunction with cruise missiles, missile defense capabilities, and supported by drone swarms, could target an adversary's radars, anti-satellite weapons, mobile missile launchers, C3I systems, and transporter-erector-launchers.[59] In the future, swarms of AI-augmented UAVs could be deployed to identify and track dispersed targets such as mobile missile launchers and suppress enemy air defenses, clearing the path for swarms of hypersonic autonomous delivery systems armed with conventional or nuclear payloads.[60] The development and deployment of offensive-dominant weapons such as hypersonic guide vehicles (HGVs), which uses boost-glide technology to propel warheads with conventional (and potentially nuclear payloads),[61] will likely compound the problem of target ambiguity and increase the risks of inadvertent escalation, thereby lowering the nuclear threshold.[62]

It is noteworthy that Chinese, US, and Russian doctrinal texts share a standard view of the potential utility of conventional hypersonic weapons to put at risk targets that hitherto only nuclear weapons could threaten, thereby bolstering their perceived strategic deterrence.[63] Moreover, in a future conflict between the US and China (or the US and Russia), *all sides* would have strong incentives to attack the other's dual-use C3I and ISR capabilities early on, and pre-emptively.[64] Chinese analysts, like their Russian counterparts, view hypersonic cruise missiles as an effective means to enhance China's nuclear deterrence posture, penetrate US missile defenses, and enhance China's offensive platform (e.g. the DF-ZF hypersonic glide platform) to support pre-emptive hypersonic operations.[65]

The maneuverability of hypersonic weapons may compound these dynamics, adding *destination ambiguity* to the destabilizing mix. In contrast to ballistic missiles, the unpredictable trajectories of hypersonic weapons will make using this weapon for signaling intent highly problematic, and potentially escalatory. Furthermore, the challenge of determining an attacker's intentions would be complicated if an adversary's dual-use ISR, early warning, or C3I systems were targeted early on in a conflict. Adversaries unable to determine the intended path or ultimate target of a 'bolt from the blue' hypersonic strike will likely assume the worst (i.e. it was in a use-it-or-lose-it situation), escalating a situation that may be intended only to signal intent. For example, Chinese analysts have expressed concern that their early warning systems would be unable to detect and counter a low signature 'bolt from the blue' hypersonic attack on its nuclear forces. This view reflects both the perceived inadequacy of Chinese early warning systems (especially vis-à-vis the US), and advances in US global prompt strike capabilities.[66]

Against the backdrop of geopolitical competition and uncertainty, the reciprocal fear of a surprise attack may heighten the risk of miscalculation, with potentially escalatory implications. Even if US Navy autonomous UUVs were patrolling the South China Seas, for example, on ISR missions, how would they convince Beijing that these capabilities would not be used for more offensive missions in the future? As demonstrated in chapter 5, the overriding operational rationale for deploying autonomous systems like UUVs in disrupted territories would be their effectiveness, precision, reliability, and intelligence gathering attributes, rather than to signal deterrence or reassurance.[67]

If Chinese early warning systems detected a hypersonic weapon launched from the US, for example, Beijing would not be sure whether China was the intended target ('destination ambiguity'). Even if it became clear that China was the intended target, Chinese leaders would still not know what assets the US intended to destroy ('target ambiguity'), or whether the weapon was nuclear or conventionally armed ('warhead ambiguity').[68] Beijing, in turn, may feel compelled to use drone swarms to track and intercept US 'dual-capable' platforms, increasing the risk of accidental and inadvertent escalation (see chapter 4).[69] Were the US to follow through with its 2018 *Nuclear Posture Review* proposal to deploy low-yield submarine-launched ballistic and cruise missiles, this strategic ambiguity would likely intensify.[70] Moreover, improvements to the Chinese missile early warning system in preparation for the adoption of a launch-under-attack nuclear posture (like Russia and the US currently maintain), mean the early detection of a US boost-guide attack would become even more critical.[71]

According to defense analyst James Acton, enabling capabilities are critical for the successful employment of hypersonic weapons.[72] In particular, military operations that require rapid decision-making (i.e. to locate, track, and accurately execute an attack) will generally place higher demands on enabling capabilities to plan and execute a strike (especially ISR) than pre-emptive or surprise attacks. To date, however, command and control, intelligence collation and analysis, and battle damage assessment systems remain undeveloped, lagging behind the progress made in hypersonic weapon technology.[73]

AI ML techniques are expected to result in significant qualitative improvements to the development of hypersonic delivery systems – and other long-range conventional and nuclear-armed precision munitions – in all of these critical enabling systems, including:[74] (1) Autonomy navigation and advanced vision-based guidance systems;[75] (2) ISR systems for targeting and tracking (especially mobile) targets; (3) missile release and sensor systems; (4) AI ML systems to decipher patterns from large data-sets and support intelligence analysis to identifying and target tracking;[76] (5) pattern interpretation to cue decision support systems for enabling 'fire and forget' missiles;[77] and (6) escalation prediction (see chapter 8).[78] For example, several states (notably China and Russia) are currently developing ML approaches to build control systems for HGVs, which cannot be operated manually because of their high velocity.[79]

These autonomous variants could also enhance hypersonic missile defenses, strengthening their resilience against non-kinetic countermeasures such as jamming, disrupting, deceiving, and spoofing.[80] Conceptually, within a matter of minutes, AI ML systems can generate a hypersonic flight plan for human review and approval and, in real-time, self-correct a missile in flight to compensate for unexpected flight conditions or a change in the target's location.[81] In theory, this AI-augmentation would enable swarms of hypersonic autonomous delivery systems to circumvent some of the remaining technical challenges that militaries face in tracking and targeting an adversary's mobile missile forces.[82] – specifically, tracking a moving target and communicating this information back to commanders in real-time, and then cue a rapid surprise or pre-emptive attack *before* the mobile launchers can be relocated.[83]

A large volume of Chinese open sources reveals prolific indigenous research into the integration of AI-powered ML techniques (especially deep neural networks), to address the technical challenges associated with the high-speed and heat-intensive re-entry dynamics of hypersonic weapons (i.e. heat control, maneuverability, stability, and targeting).[84] Chinese analysts anticipate that AI will resolve many of the intractable issues associated with HGVs' high flight envelope including: complex flight environments, severe nonlinearity, intense and rapid time-variance, and the dynamic uncertainty during the dive phase of the delivery. Further, Chinese experts broadly agree with their Western counterparts that, much like other AI-augmented strategic non-nuclear capabilities (especially drone swarms and cyber weapons), hypersonic weapons, by increasing the speed of warfare, are inherently destabilizing.

Chinese efforts to apply AI ML techniques to enhance hypersonic weapons can be understood as part of a broader strategic goal of developing 'intelligent' autonomous weapons for future multi-dimensional and multi-domain warfare.[85] Because of the multitude of intersections between hypersonic weapons with nuclear security (especially the penetrating US missile defenses), and the possibility that Chinese hypersonic weapons will carry dual-payloads[86] appreciation of the interaction between these capabilities and implications for nuclear, conventional, and cross-domain deterrence has become a critical task for analysts and policy-makers.[87]

## Conclusion

This chapter examined how AI-augmented drone swarming and hypersonic weapons might complicate missile defense, undermine states' nuclear-deterrent forces, and increase the risk of escalation. AI algorithms integrated with long-range precisions munitions, missile defense systems, and hypersonic vehicles might significantly accelerate both the speed of warfare and compress the decision-making timeframe decision-makers have to respond to a nuclear crisis. Once the remaining technical bottlenecks in deploying AI-enhanced drone swarms are surmounted, this capability will present a powerful interplay of increased range, precision, mass, coordination, intelligence, and speed in future warfare. In short, the pre-delegation of authority to autonomous systems may severely challenge the safety, resilience, and credibility of nuclear weapons in future warfare.

The exponential growth in computing performance, coupled with advances in AI ML (that can rapidly process data in real-time), will enable drones in swarms to perform increasingly complex (offensive and defensive) missions, including: searching for adversary's nuclear deterrence forces; performing nuclear-ISR operations and pre-emptive strikes; and bolstering defenses against drone swarms, hypersonic weapons, and jamming and spoofing attacks. Further, AI ML techniques are expected to make significant qualitative improvements to long-range precision munitions, including hypersonic delivery systems.

Possible strategic operations using AI-augmented drone swarms include: (1) nuclear-ISR operations to locate and track dispersed mobile missile launchers and their attendant enabling NC3 systems; (2) enhancing both existing and next-generation conventional and nuclear weapon delivery systems (e.g. ICBMs, SLBMs, and hypersonic variants); and (3) disabling or suppressing an adversary's defenses (i.e. cyberweapons or electronic warfare capabilities) as a prelude for a disarming attack. Combining speed, persistence, scope, coordination, and battlefield mass, UAS will offer states attractive asymmetric options to project military power within contested anti-access and area-denial zones.

While significant advances in power, sensor technology, and communications are needed before UAS has a game-changing strategic impact, the mere perception that nuclear capabilities face new strategic challenges may elicit distrust and exacerbate security dilemmas between nuclear-armed adversaries, in particular where strategic force asymmetries exist. Autonomous weapons such as drone swarms, perceived as low risk with ambiguous rules of engagement, and without a robust legal, normative (i.e. what constitutes a LAWS), or ethical framework, will likely become an increasingly enticing asymmetric (i.e. low-cost and relatively easy) *force majeure* to threaten a technologically superior adversary – increasing the danger of crises escalating into conflict because of misunderstanding, miscalculation, and inadvertent escalation.[88]

In sum, the uncertainties and unpredictability surrounding the deployment of accident-prone UAS operating at machine speed, coupled with competitive geopolitical pressures, could impel states to delegate increasing military control to opaque (and dual-use) algorithms. According to Paul Scharre, defense analyst at the Center for a New American Security (a US defense and security think-tank), "winning in swarm combat may depend upon having the best algorithms to enable better coordination and faster reaction times, rather than simply the best platforms."[89] How might these dynamics play out in cyberspace, and what will be the impact of AI-enhanced cyber capabilities for nuclear security? Chapter 7 picks up this challenge.

## Notes

1 Drone swarming, as a field of robotics, considers large groups of robots that, typically, operate autonomously and coordinate their behavior through decentralized command and control. The swarm, working as a collective, can, in theory, perform both simple and complex tasks in a way that a single robot would be unable to, thus increasing the robustness and flexibility of the swarm group as a whole. 'Swarming' characteristics include: (1) non-centralized autonomy; (2) ability to sense local environments' near-by swarms; and (3) ability to communicate and cooperate with other swarms to perform a specific mission. See Andrew Ilachinski, *AI, Robots, and Swarms: Issues, Questions, and Recommended Studies* (Washington, DC: Center for Naval Analysis, January 2017), p. 108; and Iñaki Navarro and Fernando Matía, "An Introduction to Swarm Robotics," *International Scholarly Research Notices* (2013), www.hindawi.com/journals/isrn/2013/608164/ (accessed February 20, 2021).

2 Recent studies generally agree that AI ML systems are an essential ingredient to enable fully autonomous systems. See Stuart Russell and Peter Norvig, *Artificial Intelligence: A Modern Approach*, 3rd ed. (Harlow: Pearson Education, 2014), p. 56; and Michael C. Horowitz, Paul Scharre, and Alexander Velez-Green, "A Stable Nuclear Future? The Impact of Autonomous Systems and Artificial

Intelligence," December 2019, *arXix*, https://arxiv.org/pdf/1912.05291.pdf (accessed March 10, 2020).

3 This chapter is derived in part from an article published in *The RUSI Journal*, April 16, 2020, copyright Taylor & Francis, available online: https://doi.org/10.1080/03071847.2020.1752026 (accessed March 10, 2020).

4 See Robert J. Bunker, *Terrorist and Insurgent Unmanned Aerial Vehicles: Use, Potentials, and Military Applications* (Carlisle, PA: Strategic Studies Institute and US Army War College Press, 2015); Zachary Kallenborn and Philipp C. Bleek, "Swarming Destruction: Drone Swarms and Chemical, Biological, Radiological, and Nuclear Weapons," *The Nonproliferation Review*, 25, 5–6 (2019), pp. 523–543; Bryan Clark, "The Emerging Era in Undersea Warfare" (Washington, DC: Center for Strategic and Budgetary Assessments, January 22, 2015); and James Johnson, "Artificial Intelligence, Drone Swarming and Escalation Risks in Future Warfare," *The RUSI Journal* (2020) www.tandfonline.com/doi/abs/10.1080/03071847.2020.1752026?journalCode=rusi20 (accessed February 20, 2021).

5 Today, robotic drone swarm technology is not fully operational. The majority of drone swarm (civilian and military) technology is still in a testing and demonstration phase. Previous (and ongoing) projects have demonstrated that swarms are capable of conducting specific (or narrow) tasks such as forming shapes, flying in formation, searching or mapping an area, patrolling a perimeter, and defending a boundary. Merel Ekelhof and Giacomo Persi Paoli, "Swarm Robotics: Technical and Operational Overview of the Next Generation of Autonomous Systems," *UNIDIR*, April 8, 2020, https://unidir.org/publication/swarm-robotics-technical-and-operational-overview-next-generation-autonomous-systems, pp. 1–2 (accessed March 10, 2020).

6 During the Cold War, for example, increased missile accuracy was generally viewed as a net positive, or natural development. See Donald A. MacKenzie, *Inventing Accuracy: Historical Sociology of Nuclear Missile Guidance* (Cambridge, MA: MIT Press, 1993).

7 The history of cruise and ballistic missile development, in particular the desire for faster and more precise missiles, demonstrates why countries might seek to accelerate the development and deployment of UAS.

8 William Knight and Karen Hao, "Never Mind Killer Robots – Here Are Six Real AI Dangers to Watch Out for in 2019," *MIT Technology Review*, January 7, 2019, www.technologyreview.com/2019/01/07/137929/never-mind-killer-robotshere-are-six-real-ai-dangers-to-watch-out-for-in-2019/ (accessed July 10, 2019).

9 The US DoD has developed directives restricting the development and use of systems with particular autonomous capabilities; 'teams' must be kept in the loop and directly make the decisions for all applications of lethal force. To date, no state has formally stated an intention to build entirely autonomous weapon systems.

10 LAMs are hybrid offensive capabilities between guided munitions and unmanned combat aerial systems. To date, the only known operational LAM is Israel's

Harop (or Harpy 2), combining a human-in-the-loop and fully autonomous mode.

11 For example, the terrorist group ISIS used remotely controlled aerial drones in its military operations in Iraq and Syria. Ben Watson, "The Drones of ISIS," *Defense One*, January 12, 2017, www.defenseone.com/technology/2017/01/drones-isis/134542/ (accessed September 10, 2019).

12 There are instances where a lone wolf drone may pose a severe threat to an F-35. For example, a UAS could be employed to place spike strips on a runway to deflate aircraft tires, deliver debris to damage jet engines, drop explosives on other targets, or even be used in a Kamikaze role during the critical takeoff or landing phases of flight, increasing the chances of damage or a catastrophic crash. Thomas S. Palmer and John P. Geis, "Defeating Small Civilian Unmanned Aerial Systems to Maintain Air Superiority," *Air & Space Power Journal*, 31, 2 (Summer 2017), pp. 102–118, p. 105.

The US, China, the UK, and France have developed and tested stealthy UAV prototypes. See Dan Gettinger, *The Drone Database* (New York: Center for the Study of the Drone, Barnard College Press, 2019).

13 For example, the Russian military reportedly deployed jammers to disrupt GPS-guided UAVs in combat zones including Syria and Eastern Ukraine. See Madison Creery, "The Russian Edge in Electronic Warfare," *Georgetown Security Studies Review*, June 26, 2019, https://georgetownsecuritystudiesreview.org/2019/06/26/the-russian-edge-in-electronic-warfare/ (accessed March 10, 2020).

14 China, the US, the UK, and France have developed and tested stealthy UAV prototypes.

15 Noah Shachtman, "Computer Virus Hits US Drone Fleet," *Wired*, July 10, 2011, www.wired.com/2011/10/virus-hits-drone-fleet/ (accessed September 10, 2019).

16 AI-infused algorithms that can integrate sensor information, consolidate targeting, automate maintenance, and process navigation data information are currently being developed and tested in anticipation of the kind of high-intensity future threat environments posed by drone swarming. To date, however, small drone technology does not enable drones to fly at speeds where they could be or remain close to the aircraft. The majority of concepts today involve either medium-sized (i.e. MQ-9) drones acting as wingmen to a manned fighter jet (i.e. F-35), or small drones released as a payload that does not remain near the fighter. Kris Osborn, "The F-35 Stealth Fighter: The Safest Fighter Jet Ever Made?" *The National Interest*, September 27, 2019, https://nationalinterest.org/blog/buzz/f-35-stealth-fighter-safest-fighter-jet-ever-made-83921 (accessed March 10, 2020).

17 These 'drone wingmen' aircraft can direct the pilot to a potential target and alert them to incoming threats, among other capabilities. Several US DoD programs, including the US Air Force Research Laboratories, Have Raider, Skyborg, and Mako Unmanned Tactical Aerial Platform (UTAP-22), are developing these kinds of manned-unmanned teaming systems. See Douglas Birkey, David

Deptula, and Lawrence Stutzriem, "Manned-Unmanned Aircraft Teaming: Taking Combat Airpower to the Next Level," Mitchell Institute Policy Papers, Vol. 15, July 2018, http://docs.wixstatic.com/ugd/a2dd91_65dcf7607f144e729dfb1d873e1f0163.pdf (accessed 10 February 2019).
18 The US Air Force's high-powered microwave system called Tactical High-Power Microwave Operational Responder (THOR), for example, is designed to protect bases against swarms of drones. Andrew Liptak, "The US Air Force has a New Weapon Called THOR that Can Take Out Swarms of Drones," theverge, June 21, 2019, www.theverge.com/2019/6/21/18701267/us-air-force-thor-new-weapon-drone-swarms (accessed September 10, 2019).
19 Quote from an interview in *The Washington Post Live*, "David Ignatius and Pentagon's Robert Work Talk About New Technologies to Deter War," *Washington Post*, March 30, 2016, www.washingtonpost.com/blogs/post-live/wp/2016/02/29/securing-tomorrow-with-david-ignatius-whats-at-stake-for-the-world-in-2016-and-beyond/ (accessed September 10, 2019).
20 UAVs used in swarming operations do not necessarily need to be 'fully-autonomous;' humans could still decide to execute a lethal attack.
21 Patricia Lewis, Heather Williams, Benoit Pelopidas, and Susan Aghlani, *Too Close for Comfort: Cases of Near Nuclear Use and Options for Policy* (London: Chatham House Report, Royal Institute of International Affairs, 2014).
22 The Russian military also plans to augment 'AI killer robots' into unmanned aerial and undersea vehicles for swarming missions. For example, Kalashnikov, a Russian defense contractor, has reportedly built an unmanned ground vehicle (the *Soratnik*) and plans to develop a broad range of autonomous systems infused with advanced AI machine learning algorithms. Tristan Greene, "Russia is Developing AI Missiles to Dominate the New Arms Race," *The Next Web*, July 27, 2017, https://thenextweb.com/artificial-intelligence/2017/07/27/russia-is-developing-ai-missiles-to-dominate-the-new-arms-race/ (accessed March 10, 2020).
23 Elsa Kania, "China's Strategic Ambiguity and Shifting Approach to Lethal Autonomous Weapon Systems," *Lawfare*, April 17, 2018, www.lawfareblog.com/chinas-strategic-ambiguity-and-shifting-approach-lethal-autonomous-weapons-systems (accessed March 10, 2020).
24 Lin Juanjuan, Zhang Yuantao, and Wang Wei, "Military Intelligence is Profoundly Affecting Future Operations," *Ministry of National Defense of the People's Republic of China*, September 10, 2019, www.mod.gov.cn/jmsd/2019-09/10/content_4850148.htm (accessed September 12, 2019).
25 Like AI applications more broadly, there are many ethical and moral issues associated with autonomous weapon systems that are worthy of discussion. For example, see Peter Asaro, "On Banning Lethal Autonomous Weapon Systems: Human Rights, Automation, and the Dehumanization of Lethal Decision-Making," *International Review of the Red Cross*, 94, 886 (2012), pp. 687–709; Heather M. Roff, "Meaningful Human Control or Appropriate Human Judgment? The Necessary Limits on Autonomous Weapons," Arizona State University Global Security Initiative Briefing Paper 2016, https://globalsecurity.

asu.edu/sites/default/files/files/Control-or-Judgment-Understanding-the-Scope.pdf (accessed March 10, 2020); United Nations Institute for Disarmament Research, "The Weaponization of Increasingly Autonomous Technologies: Considering Ethics and Social Values," No. 3 (2015), www.unidir.org/files/publications/pdfs/considering-ethics-and-social-values-en-624.pdf (accessed September 10, 2019); and Michael C. Horowitz, "The Ethics and Morality of Robotic Warfare: Assessing The Debate Over Autonomous Weapons," *Daedalus*, 145, 4 (2016), pp. 25–36.

26 Modeling interactions with other agents (especially humans) in either a competitive or a collaborative context is inherently problematic because human behavior is often unpredictable. Ilachinski, *AI, Robots, and Swarms*, p. xv.

27 "Autonomy in Weapon Systems," US Department of Defense, *Directive Number 3000.09*, May 8, 2017, www.esd.whs.mil/Portals/54/Documents/DD/issuances/dodd/300009p.pdf (accessed September 10, 2019); and Lin, Zhang, and Wang, "Military Intelligence is Profoundly Affecting Future Operations."

28 A notable exception is Arthur H. Michel, *Unarmed and Dangerous: The Lethal Application of Non-Weaponized Drones* (Washington, DC: The Center for the Study of the Drone at Bard College, March 2020).

29 To bring to bear the full potential of drone swarming and, at the same time, allow for appropriate levels of human involvement, some analysts believe that swarms will require new command and control mechanisms and systems. Conversely, another camp argues that human participation is contradictory to a swarm. Because a swarm is inherently unpredictable, humans would be unable to control its behavior in a way that is appropriate or meaningful. See Paul Scharre, *Robotics on the Battlefield – Part II: The Coming Swarm* (Washington, DC: Center for a New American Security, 2014).

30 Many types of UAS could be used for these types of operations, including USVs, UUVs, and UAVs. Between 2011 and 2017, the Predator and Reaper UAVs performed circa 127,390 combat ISR operations – open-source information does not specify how many of those missions resulted in actual strikes. Ibid., p. 9.

31 Joseph Trevithick, "Navy's Sea Hunter Drone Ship has Sailed Autonomously to Hawaii and Back Amid Talk of the New Roles," *The Drive*, February 4, 2019, www.thedrive.com/the-war-zone/26319/usns-sea-hunter-drone-ship-has-sailed-autonomously-to-hawaii-and-back-amid-talk-of-new-roles (accessed September 10, 2019).

32 Elias Groll, "How AI Could Destabilize Nuclear Deterrence," *Foreign Policy*, April 24, 2018, https://foreignpolicy.com/2018/04/24/how-ai-could-destabilize-nuclear-deterrence/ (accessed September 10, 2019).

33 Given the current limits on drone range (i.e. battery power) and limited payload, it is unlikely that drone technology will mature sufficiently to represent a credible threat to states' nuclear assets (or other hardened targets) in the near-term (i.e. within five years) – unless, for example, UAVs can infiltrate hardened targets via an air duct or other like passage. James S. Johnson, "Artificial Intelligence: A Threat to Strategic Stability," *Strategic Studies Quarterly*, 14, 1 (2020), pp. 16–39.

34 Johnson, "Artificial Intelligence: A Threat to Strategic Stability," pp. 17–19.
35 Rebecca Hersman, Reja Younis, Bryce Farabaugh, Bethany Goldblum, and Andrew Reddie, *Under the Nuclear Shadow: Situational Awareness Technology & Crisis Decision-making* (Washington, DC: The Center for Strategic and International Studies, March 2020).
36 Tom Simonite, "Moore's Law is Dead. Now What?" *MIT Technology Review*, May 13, 2016, www.technologyreview.com/ (accessed 10 February 2019). In addition to UAVs, emerging space technologies will soon enable drone-like surveillance from space incorporating similar ML techniques. Larger satellite constellations coupled with smaller individual satellites are expected to provide continuous coverage over large geographical ranges.
37 The value of drones in these scenarios does not mean that they are the *only* or necessarily most effective way to fulfill these missions. Jonathan Gates, "Is the SSBN Deterrent Vulnerable to Autonomous Drones?" *The RUSI Journal*, 161, 6 (2016), pp. 28–35.
38 In 2011, students at the Massachusetts Institute of Technology (MIT) presented the fully autonomous, fixed-wing *Perdix* UAV capable of between-drone communication at the 2011 Air Vehicle Survivability Workshop. In addition to the US, Russia, South Korea, and China are also actively pursuing drone swarm technology programs. Kallenborn and Bleek, "Swarming Destruction: Drone Swarms and Chemical, Biological, Radiological, and Nuclear Weapons," pp. 1–2.
39 For example, the US DoD has invested heavily in developing software to detect suspicious behavior from drone aerial footage automatically. Michel *Unarmed and Dangerous: The Lethal Application of Non-Weaponized Drones*, p. 13.
40 In April 2018, for example, the US drone manufacturer AeroVironment demonstrated an automatic target hand-off sequence between a hand-launched surveillance drone (RQ-20B Puma), and a Switchblade loitering munition, significantly reducing the length of the targeting cycle against a fast-moving target. Press Release, "AeroVironment Successfully Conducts Maritime Demonstration of Puma-Switchblade Automated Sensor-to-Shooter Capability to Counter Swarm Attacks," *National Harbor: AeroVironment*, April 9, 2018, www.avinc.com/resources/press-releases/view/aerovironment-successfully-conducts-maritime-demonstration-of-puma-switchbl (accessed 10 February 2019).
41 Currently, F-35s and UAVs are unable to penetrate in ranges much beyond China's periphery; it is believed that Chinese strategic assets are not positioned on China's coastline.
42 In 2015, Russia revealed the development of a large nuclear-armed UUV delivery vehicle, *Poseidon* (also known as Status-6). The US is also developing an 'optionally manned' nuclear-capable long-range bomber, the B-21 *Raider*, which could potentially carry nuclear payloads. Other unmanned combat aerial vehicles prototypes (e.g. Northrop Grumman X-47B, the Dassault nEUROn, and the BAE Systems Taranis) could also feasibly be used in nuclear attacks. Ria Novosti, "Russia Could Deploy Unmanned Bomber After 2040 – Air Force," *GlobalSecurity.org*, February 8, 2012, www.globalsecurity.org/wmd/library/

news/russia/2012/russia-120802-rianovosti01.htm (accessed February 10, 2019); and Robert M. Gates, "Statement on Department Budget and Efficiencies" (US Department of Defense, January 6, 2011), http://archive.defense.gov/Speeches/Speech.aspx?SpeechID=1527 (accessed September 10, 2019).
43 While similar vulnerabilities exist in other technically advanced weapon systems discussed in this chapter, UAV swarms require a high degree of autonomy, making them more susceptible to these kinds of attacks.
44 Drone swarms could be programmed to make regular changes to their route to counter an adversary using AI ML-augmented intelligence to anticipate its trajectory and defeating detection. Kallenborn and Bleek, "Swarming Destruction: Drone Swarms and Chemical, Biological, Radiological, and Nuclear Weapons," p. 16.
45 Pawlyk Oriana, "Pentagon Still Questioning How Smart to Make its Drone Swarms," *Military.Com*, February 7, 2019, www.military.com/defensetech/2019/02/07/pentagon-still-questioning-how-smart-make-its-drone-swarms.html (accessed September 10, 2019).
46 Mike Pietrucha, "The Need for SEAD Part 1: The Nature of SEAD," *War on the Rocks*, May 17, 2016, https://warontherocks.com/2016/05/the-need-for-sead-part-i-the-nature-of-sead/ (accessed September 10, 2019).
47 Polat Cevik, Ibrahim Kocaman, Abdullah S. Akgul, and Barbaros Akca, "The Small and Silent Force Multiplier: A Swarm UAV-Electronic Attack," *Journal of Intelligent and Robotic Systems*, 70 (April 2013), pp. 595–608.
48 Liu Xuanzun. "Chinese Helicopter Drones Capable of Intelligent Swarm Attacks," *Global Times*, May 9, 2019, www.globaltimes.cn/content/1149168.shtml (accessed September 10, 2019).
49 Arthur H. Michel, *Unarmed and Dangerous: The Lethal Application of Non-Weaponized Drones*, p. 17.
50 For example, DARPA's Anti-Submarine Warfare Continuous Trail Unmanned Vessel program, designed to track quiet diesel-electric submarines with USVs from the surface (see chapter 5).
51 Today, all operable air-defense systems operate under human supervision, and fully automatic mode is generally used to defend against anti-ship cruise missiles. See John K. Hawly, "Patriot Wars: Automation and the Patriot Air and Missile Defense Systems" (Washington, DC: CNAS, January 2017).
52 Swarms could be used as decoys to create false signatures (e.g. time delays in tricking a defender), or to force an enemy to reveal (or 'light up') its weapons; by switching on their radars to attack drone swarms. The idea of drones in swarms waiting for aircraft to collide with them, or drones homing in on aircraft and scoring a kamikaze-like attack, is analogous to the way hydrogen balloons were employed as obstacles during World War II – a tactic known in military doctrine as 'barrage defense.' Leslie F. Hauck and John P. Geis II, "Air Mines: Countering the Drone Threat to Aircraft," *Air & Space Power Journal*, 31, 1 (Spring 2017), pp. 26–40.
53 For example, researchers at China's Tianjin University have conducted studies on path planning and software to coordinate swarms of underwater gliders

working together. Shijie Liu et al., "Cooperative Path Planning for Networked Gliders under Weak Communication," *Proceedings of the International Conference on Underwater Networks & Systems*, Article No. 5, 2014.

54 While the Missile Defense Agency (MDA) is developing lasers for drones, the size of a drone needed to power a laser of meaningful power would be substantial. Thus, the likelihood we will see this on drones in the near-term is considered low. The MDA estimates that the prototype laser for a fighter-sized platform will likely be completed by the end of 2023. The MDA recently requested a significant budget to develop a drone-mounted laser program. Jen Judon, "MDA awards contracts for a drone-based laser design," *Defense News*, December 11, 2017, www.defensenews.com/land/2017/12/11/mda-awards-three-contracts-to-design-uav-based-laser/ (accessed September 10, 2019).

55 Ekelhof and Paoli, "Swarm Robotics: Technical and Operational Overview of the Next Generation of Autonomous Systems," p. 53.

56 Supplying sufficient power for swarms of UAVs (or UUVs) for extended periods would require significant improvements in either battery technology, air-independent propulsion, or fuel-cell technology. It may also require the development of some form of energy storage mechanism that has yet to be envisaged. Gates, "Is the SSBN Deterrent Vulnerable to Autonomous Drones?"

57 As technology advances – particularly AI – an increasing number of states will likely consider developing and operating LAWS. The international community considers the implications of LAWS in conversations held under the auspices of the United Nations Convention on Certain Conventional Weapons, a multilateral arms control agreement to which the US became a party in 1982. The US and Russia (albeit for different reasons) have consistently opposed banning LAWS. China supports a ban on the use, but not development, of LAWS, which it defines to be indiscriminate – lethal systems that do not have any human oversight and cannot be terminated. However, some have argued that China is maintaining 'strategic ambiguity' about its position. See Kelly M. Sayler and Michael Moddie, "International Discussions Concerning Lethal Autonomous Weapon Systems," Congressional Research Service (In Focus), October 15, 2020, https://fas.org/sgp/crs/weapons/IF11294.pdf (accessed March 10, 2020).

58 This probing tactic would likely only be effective if *both* sides possessed these autonomous systems. If the receiver of an attack with these systems did not have these capacities, they would likely develop counterstrategies and capabilities to mitigate the asymmetric advantages of these weapons. For a discussion of the impact of the arms race and effect on arms racing dynamics and military drones, see Michael J. Boyle, "The Race for Drones," *Orbis*, November 24 (2014), pp. 76–94.

59 James M. Acton, *Silver Bullet? Asking the Right Questions About Conventional Prompt Global Strike* (Washington, DC: Carnegie Endowment for International Peace, 2013).

60 Currently, ballistic missiles mounted with hypersonic guide vehicles can only maneuver while inside the atmosphere, and the atmosphere's density at the

turning point dictates their turn rate. Tight turns are only possible near the ground and close to the target.
61 Russia, China, and the US have been the most active states in the development of hypersonic weapons. To date, however, no state has emerged as the dominant leader in this nascent technology. See James M. Acton (ed.), with Li Bin, Alexey Arbatov, Petr Topychkanov, and Zhao Tong, *Entanglement: Russian and Chinese Perspectives on Non-Nuclear Weapons and Nuclear Risks* (Washington, DC: Carnegie Endowment for International Peace, 2017), p. 54.
62 Currently, the drag associated with HGVs remaining in the atmosphere will require new propulsion technologies and innovation in ablative materials to absorb the increased heat, neither of which are expected to emerge in the near-term. The author would like to thank an anonymous reviewer for making this point.
63 A particular concern identified by Russian and Chinese analysts is that a combination of US ballistic missile defense and high-precision conventional weapons (such as hypersonic weapons) could permit the US to attempt a disarming first strike without crossing the nuclear Rubicon.
64 Avery Goldstein, "First Things First: The Pressing Danger of Crisis Instability in US–China Relations," *International Security*, 37, 4 (Spring 2013), pp. 67–68.
65 According to the 2019 US DoD Missile Defense Review: "Russia and China are developing advanced cruise missiles and hypersonic missile capabilities that can travel at exceptional speeds with unpredictable flight paths that challenge our existing defensive systems," Office of the Secretary of Defense, *2019 Missile Defense Review*, https://media.defense.gov/2019/Jan/17/2002080666/-1/-1/1/2019-MISSILE-DEFENSE-REVIEW.PDF (accessed March 10, 2020).
66 Lora Saalman, "Fear of False Negatives: AI and China's Nuclear Posture," *Bulletin of the Atomic Scientists*, April 24, 2018, https://thebulletin.org/2018/04/fear-of-false-negatives-ai-and-chinas-nuclear-posture/ (accessed March 10, 2020).
67 Paul Scharre, "Highlighting Artificial Intelligence: An Interview with Paul Scharre," *Strategic Studies Quarterly*, 11, 4 (November 2017), pp. 18–19.
68 The US is not currently known to be pursuing hypersonic delivery systems with nuclear warheads, so this warhead ambiguity should be less of a problem. It is less clear, however, whether Russia and China are developing hypersonic weapons with nuclear payload optionally.
69 The term 'dual-capable' rather than 'dual-use' is used in this context because while many weapon systems can carry either nuclear or conventional warheads (including hypersonic boost vehicles), only a smaller number of these dual-use weapons are assigned *both* nuclear and conventional missions.
70 US Department of Defense, *Nuclear Posture Review* (Washington, DC: US Department of Defense, February 2018), pp. 54–55.
71 Analysts have noted that there are calls within China to adopt a launch-on-warning strategy and develop the technology to enable this capability. See Acton, *Entanglement: Russian and Chinese Perspectives on Non-Nuclear Weapons and Nuclear Risks*, p. 79.

72 Acton, *Silver Bullet? Asking the Right Questions About Conventional Prompt Global Strike*, p. xiv.
73 Assessing the precise status of US enabling capabilities is challenging because they are so highly classified. Ibid., pp. 88–90.
74 Because ICBMs and SLBMs rely on automation to set their flight trajectory and navigate to their target, they operate *de facto* autonomously once launched. Thus, while autonomy enhances the strategic value of missile delivery systems, it is not an operational prerequisite, except perhaps in the under-water domain, where munitions cannot be easily operated remotely. See Vincent Boulanin and Maaike Verbruggen, *Mapping the Development of Autonomy in Weapon Systems* (Stockholm: SIPRI, 2017), p. 57.
75 Existing navigation systems tend to rely heavily on pre-mapping and navigate autonomously to identify paths and obstacles; however, navigation systems will need to incorporate advanced vision-based guidance in-built pre-mapping systems. Advances in machine learning techniques could significantly improve the vision-based guidance systems of these sub-systems and enable autonomy. Ibid., p. 114.
76 For example, the DoD Defense Innovation Unit plans to partner with the Joint Artificial Intelligence Center to mine large data-sets across multiple aircraft platforms and ground vehicles to develop analytics and predictive maintenance applications for the US Air Force and Army.
77 So-called 'fire-and-forget' (or semi-autonomous) missiles allow the onboard sensors and computer to guide a missile to its target without further operator communications following initial target selection and fire authorization.
78 Chinese analysts have begun research into the use of big-data and deep-learning AI techniques to enhance the processing speed and intelligence analysis of satellite images to support the military early warning capabilities, enabling a 'prediction revolution' in future warfare.
79 For example, Chinese analysts claim that future cruise missiles will be fused with AI and autonomy, enabling commanders to control them in real-time or deploy them in fire-and-forget operations. Zhao Lei, "Nation's Next Generation of Missiles to be Highly Flexible," *China Daily*, August 16, 2016, www.chinadaily.com.cn/china/2016-08/19/content_26530461.htm (accessed September 10, 2019).
80 Vincent Boulanin (ed.), *The Impact of Artificial Intelligence on Strategic Stability and Nuclear Risk Vol. I Euro-Atlantic Perspectives* (Stockholm: SIPRI Publications, May 2019), p. 56.
81 US government-funded Sandia National Laboratories, which has made and tested hypersonic vehicles for more than thirty years, recently established an academic research coalition, 'Autonomy New Mexico,' whose mission is to create artificially intelligent aerospace systems. Bioengineer, "Future Hypersonics Could Be Artificially Intelligent," *Bioengineer.org*, April 18, 2019, https://bioengineer.org/future-hypersonics-could-be-artificially-intelligent/ (accessed March 10, 2020).
82 Technical challenges remain in AI's ability to autonomously detect and cue precision missile munitions, especially in cluttered and complex environments.

This weakness is caused partly by the poorly understood nature of AI's ability to mimic human vision and cognition. David Deutsch, "Creative Blocks," *Aeon* October 3, 2012, https://aeon.co/essays/how-close-are-we-to-creating-artificial-intelligence (accessed September 10, 2019).

83 Long and Green, "Stalking the Secure Second Strike: Intelligence, Counterforce, and Nuclear Strategy," pp. 21–24.

84 Researchers from China's People's Liberation Army, College of Mechatronic Engineering and Automation of the National University of Defense Technology, Harbin University, and Beijing Institute of Tracking and Telecommunications Technology have collaborated to address the technical challenges faced in control dynamics with HGVs.

85 For example, drone swarms, robotics, precision guidance munitions, early warning systems, and cyber and electronic warfare capabilities.

86 China is actively exploring the use of AI to improve targeting and the precision of missile guidance systems – of conventional and, possibly, nuclear weapons. While technically feasible, the US does not currently see any role for unmanned bombers in nuclear weapon delivery. See "2017 Target Recognition and Artificial Intelligence Forum," April 22, 2017, www.csoe.org.cn/meeting/trai2017/ (accessed September 10, 2019).

87 US Department of Defense. *Military Defense Review 2019* (Washington, DC: Office of the Secretary of Defense, 2019), www.defense.gov/Portals/1/Interactive/2018/11-2019-Missile-Defense-Review/The%202019%20MDR_Executive%20Summary.pdf (accessed September 10, 2019).

88 Research and Development costs to create drones able to process data for ISR missions, perform intra-swarm communication, and navigate complex battlefield scenarios can be relatively expensive. Michael Safi, "Are Drone Swarms the Future of Aerial Warfare?" *The Guardian*, December 4, 2019, www.theguardian.com/news/2019/dec/04/are-drone-swarmsthe-future-of-aerial-warfare (accessed September 10, 2019).

89 Paul Scharre, "Counter-Swarm: A Guide to Defeating Robotic Swarms," *War on the Rocks*, March 31, 2015, http://warontherocks.com/2015/03/counter-swarm-a-guide-to-defeatingrobotic-swarms/ (accessed September 10, 2019).

# 7

# The AI-cyber security nexus

How might AI-infused cyber capabilities be used to subvert, or otherwise compromise, the reliability, control, and use of states' nuclear forces? This chapter argues that AI-enhanced cyber capabilities could increase the risk of inadvertent escalation caused by the co-mingling of nuclear and non-nuclear weapons, and the increasing speed of warfare.[1] It examines the potential implications of cyber (offensive and defensive) capabilities augmented with AI applications for nuclear security. The chapter finds that future iterations of AI-enhanced cyber counterforce capabilities will complicate the cyber-defense challenge, thereby increasing the escalatory effects of offensive cyber capabilities.

Several US national security officials believe that AI, used as force multipliers for both defensive and offensive cyber weapons, will have a transformative impact on cybersecurity.[2] Former Director of US National Intelligence Daniel Coats recently warned that AI could increase US vulnerability to cyber-attacks, weaken its ability to attribute such attacks, improve the effectiveness and capabilities of adversaries' weapon and intelligence systems, and cause accidents, and related liability issues.[3] To be sure, the line between AI cyber-offense and cyber-defense will likely remain an obscure one.[4] Bernard Brodie's words of caution about the advent of nuclear weapons six decades ago resonate: "The [military] bias towards the offensive creates special problems in any *technologically new situation* where there is little or no relevant war experience to help one to reach a balanced judgment" (emphasis added).[5]

On the one hand, AI might reduce a military's vulnerability to cyber-attacks. AI cyber-defense tools (or 'counter-AI'), designed to recognize changes to patterns of behavior and anomalies in a network, and automatically identify software code vulnerabilities,[6] could form a more robust defense against cyber subversions.[7] For example, if individual code fragments mimic existing malware structures, ML algorithms could locate vital evidence to ascertain an attacker's identity.[8] According to the Pentagon's new AI strategic report, "AI can enhance our ability to predict, identify, and respond to cyber and physical threats from a range of sources."[9] Besides, the DoD's Defense Innovation Unit is prototyping an application – related to Project VOLTRON – that leverages AI to decipher high-level strategic questions, map probabilistic chains of events, and develop alternative strategies, in order to make DoD systems more resilient to AI-augmented cyber-attacks and configure and fix errors more quickly than humans.[10]

On the other hand, autonomy itself might increase a military's vulnerability to cyber-attacks. AI will likely increase the anonymity of the attack in cyberspace, which is particularly amenable to stealth, obscuration, deception, and stratagem.[11] For example, an adversary could use malware to take control, manipulate, or fool the behavior and pattern recognition systems of autonomous systems, such as the DoD's Project Maven. Offensive attacks such as this would be relatively easy to execute but very difficult to detect, attribute, or effectively counter. For example, analysts found it extremely difficult to correctly identify a malware that infected systems control of US UAVs' cockpits (Predator and Reaper drones) during missions over Middle-Eastern warzones US Creech Air Force Base in Nevada in 2011.[12] Ironically, the use of ML to strengthen cybersecurity may increase the points at which an attacker can interact, and thus potentially manipulate, or otherwise interfere, with a network.

This problem-set is compounded by the lack of an agreed framework or understanding of what constitutes escalatory behavior (or 'firebreaks') in cyberspace.[13] Thus, a cyber operation intended as a signal (i.e. coercive diplomacy) could go undetected by the target, or worse, misinterpreted as an offensive attack. Even if information relating to an operation of this kind is accurately identified on time, the motives behind them remain ambiguous or misperceived. According to Robert Jervis, "it is likely that the country that is the object of the attack would assume that any effect was the intended one."[14]

Most observers acknowledge that no strategy combining offense and defense cyber operations could realistically be expected to deter *all* malign cyber-attacks.[15] While AI-enhanced cyber capabilities can enhance deterrence, they can simultaneously incentivize others to attack, thereby exacerbating the paradox of *enhanced capabilities* and *increased vulnerabilities* in the cyber domain.[16] As the historical record demonstrates, this capability-vulnerability paradox is heightened when states are both dependent on a particular capability – such as AI and cyber tools. Their access to and ability to use the capabilities is vulnerable to an adversary's exploitation or subversion, creating first-mover incentives.[17]

This chapter is organized into three sections. The first examines how AI applications designed to enhance cybersecurity for nuclear forces could simultaneously make cyber-dependent nuclear weapon systems more vulnerable to cyber-attacks. It argues that the intense time pressures that would loom large with the decision to use nuclear weapons mean that the threat of AI-enhanced cyber offense would be difficult to detect, authenticate, or attribute within the short timeframe for initiating a nuclear strike. The section addresses this question with two hypothetical scenarios.

The second section investigates how rapid advances in AI and increasing military autonomy might amplify the speed, power, and scale of future attacks in cyberspace – especially targeting an adversary's command and control systems. It describes the critical intersections between AI systems and cybersecurity that are most relevant to the military domain. This section also describes the new offensive concepts that AI machine learning might enable (i.e. left-of-launch operations), and how the fear of pre-emptive attacks could cause use-them-or-lose-them situations in cyberspace.[18] The final section examines how advances in AI machine learning (ML) techniques could manipulate the digital information landscape, creating new escalation pathways where decisions about nuclear weapons take place.

## Cybersecurity and nuclear weapons

At a strategic level of conflict, AI applications designed to enhance cybersecurity for nuclear forces could simultaneously make cyber-dependent nuclear weapon systems (i.e. communications, data processing, or early warning sensors) more vulnerable to cyber-attacks.[19] Today, it is thought possible that a cyber-attack (i.e. spoofing, hacking, manipulation, and digital jamming) could infiltrate a nuclear weapon system, threaten the integrity of its communications, and ultimately (and possibly unbeknown to its target) gain control of its – possibly dual-use – command and control systems (see chapter 8).[20] Because of the intense time pressures associated with the decision to use nuclear weapons – especially where a state maintains a launch-on-warning posture – AI-enhanced cyber-attacks against nuclear systems would be almost impossible to detect and authenticate. Further, warning signals would be difficult to authenticate, let alone attribute, within the short timeframe for initiating a nuclear strike.

A shared concern of China, the US, and Russia (albeit with varying degrees of sensitivity) are the potential threats posed by AI-augmented cyber warfare that might impel states to adopt (or rely more heavily upon) a launch-on-warning nuclear posture, or a policy of pre-emption during a crisis.[21] In short, conventional strikes against an adversary's nuclear-deterrent forces (especially command, control, communications, and intelligence (C3I) systems) combining AI-enabled intelligence, surveillance, and reconnaissance (ISR) and cyber capabilities, would amplify the potentially destabilizing impact of such an operation.[22]

Advances in AI could also exacerbate this cybersecurity challenge by enabling improvements to the cyber offense. By automating advanced persistent threat (APT) operations, ML and AI could dramatically reduce the extensive resources and skill required to execute APT operations (or 'hunting for weaknesses'), especially against hardened nuclear targets.[23] The machine speed of AI-augmented cyber tools could enable an attacker to exploit a narrow window of opportunity to penetrate an adversary's cyber-defenses or use APT tools to find new vulnerabilities faster and more easily than before. As former US Chairman of the Joint Chiefs, General Joseph Dunford, recently warned, "the *accelerated speed of war* ensures the ability to recover from early missteps is greatly reduced" (emphasis added).[24] For example, when docked for maintenance, air-gapped nuclear-powered submarines, considered secure when submerged, could become increasingly vulnerable to a new generation of low-cost – possibly black-market – and highly automated APT cyber-attacks.[25]

An attacker could also apply AI ML techniques to target autonomous dual-use early warning and other operating systems (e.g. C3I, ISR, early warning, and robotic control networks) with 'weaponized software' such as hacking, subverting, spoofing, or tricking, causing unpredictable and potentially undetectable errors, malfunctions, and behavioral manipulation to weapon systems – or 'data-poisoning.'[26] For example, an attacker could poison a data-set to inhibit an algorithm from learning specific patterns, or to insert a secret backdoor that can be used to trick the system in the future.[27]

Moreover, as the linkages between digital and physical systems (or the internet of things (IoT))[28] expand, the potential for an adversary to use cyber-attacks in both kinetic and non-kinetic attacks will increase.[29] Former US Director of National Intelligence, James Clapper, warned in 2016 that the IoT would further enable state proxy actors to monitor, track, and target individuals in espionage operations.[30] A significant risk variable in the operation of autonomous systems is the time that passes between a system failure (i.e. performing in a manner other than how the human operator intended), and the time it takes for a human operator to take corrective action. If the system failure is the result of a deliberate act, this timeframe is compressed.[31]

How could AI-augmented cyber capabilities create new pathways for accidental or inadvertent escalation? To illustrate these dynamics:[32] State A launches a malicious AI-enhanced cyber-attack to spoof State B's AI-enabled autonomous sensor platforms and automated target recognition systems, in such a way that the weapon system (e.g. a human-supervised ATR system) is fooled into interpreting a civilian object (e.g. an airliner) as a military target.[33] B, in response, based on manipulated or erroneous information, and the inability of human supervisors to detect the spoofed imagery that fooled the weapons' automated target recognition algorithm in time to take corrective action, accidentally (and unintentionally) escalates a situation.[34]

In this example, the spoofing attack on the weapon system's algorithm is executed in such a way that the imagery appears to the recognition system as indistinguishable from a valid military target;[35] escalating a situation based on a false premise that would be unlikely to fool the human eye.[36] Failure to fully comprehend or anticipate the consequences of the complexity caused by the interplay of human–machine interactions in cyberspace could result in confusion and unintended and accidental escalation to a strategic level, for example, cyber (defensive or offensive) operations against adversaries dual-use C3I systems.[37]

Furthermore, the explainability (or 'black box') problem associated with AI might compound these potential escalation dynamics.[38] Insufficient understanding of how and why AI algorithms reach a particular judgment or decision might complicate the task of determining if data-sets had been deliberately compromised to produce false outcomes (e.g. attacking incorrect targets or misdirecting allies during combat).[39] Moreover, as humans and AI team up to accomplish particular missions, the opacity associated with how AI systems reach a decision may cause an operator to have either too much or too little confidence in a system's performance.[40] The lack of understanding in this evolving human–machine information interaction, coupled with increasingly advanced ML algorithms and military dependency on AI and automation, will likely lead to unpredictable system behavior and unexpected outcomes – chapter 8 will return to this idea.

As explained in chapter 1, unless a system's ML algorithm is terminated at the end of the training phase, once deployed, it could potentially learn something it was not intended to, or even perform a task or mission that its human designers do not expect it to do.[41] In sum, technologies that improve the reliability and speed of information processed and disseminated by early warning systems might paradoxically also increase the vulnerabilities of these networks, creating new first-mover advantages and escalation pathways during a crisis that spark unintentional (or intentional) warfare.[42] Put another way; AI might be developed to enable a state to pursue a predetermined escalatory path. Thus, increased escalation risk as a result of technology is *not always* inadvertent or accidental.[43]

## AI 'cyber guns': a losing game of Whac-A-Mole?

US Cyber Fleet Command Commander, Michael Gilday, recently told the Senate Armed Services Committee that the US Navy must "improve an ability to *detect new and unknown malware* proactively ... so we [the US] can act quickly using advanced analytics enabled by AI and ML," which may give the US a "tactical advantage" to identify malicious activity early on (emphasis added).[44] Even if analysts can obtain high-quality and reliable intelligence, they may not want to reveal it, because doing so could compromise a source, capability, or tactic.[45]

Moreover, most observers now acknowledge that no strategy (i.e. combining offense and defense cyber operations) could realistically be expected to deter *all* malign cyber-attacks. While AI-enhanced cyber capabilities can enhance deterrence, they can simultaneously incentivize others to attack, thus, exacerbating the paradox of *enhanced capabilities* (and system redundancy) and *increased vulnerabilities* in the cyber domain. As the historical record attests, this capability-vulnerability paradox is heightened when states are both dependent on a particular capability (such as AI and cyber tools). Their access to and ability to use the capabilities is vulnerable to an adversary's exploitation or subversion, creating first-mover incentives.[46]

Rapid advances in AI and military autonomy will likely amplify the ability to use cyber-offense tools to manipulate and subvert the digital landscape with increasing potential speed, power, and scale.[47] In response to these anticipated vulnerabilities, and to achieve the first-mover advantages, the increased speed in AI-augmented cyber tools could confer; China, Russia, and the US have continued to harden their AI cyber-defenses.[48] Despite these parallel efforts, divergent views exist between how strategic communities perceive the nature of the risks posed by AI-enhanced cyber-attacks against nuclear (and dual-use) C3I systems. Thus, during a crisis, AI-augmented cyber intelligence gathering tools targeting an adversary's command and control assets may be misinterpreted as a prelude for an imminent pre-emptive attack on its nuclear forces, undermining strategic stability.[49]

There are three significant intersections between AI systems and cyber-security that are most salient to the military domain. First, advances in autonomy and ML techniques mean that a much broader range of physical systems is now vulnerable to cyber-attacks (e.g. hacking and data-poisoning).[50] Second, cyber-attacks on AI systems can offer attackers access to ML algorithms, trained models being used by the application, and potentially vast amounts of data from facial recognition and intelligence collection and analysis systems (e.g. satellite navigation and imagery systems used to cue precision munitions and supported ISR missions).

Third, the use of AI systems in conjunction with existing cyber offense tools will enable sophisticated cyber-attacks to be executed at a larger scale – both geographically and across networks – at faster speeds and simultaneously across multiple military domains with improved anonymity. AI ML might also enable new offensive concepts such as 'left-of-launch operation,'[51] thereby compounding the ambiguities and fear about pre-emptive attacks in cyberspace and causing use-them-or-lose-them situations.[52] In sum, despite the relatively benign augmentation mechanisms used to develop cyber offensive capabilities with AI systems, the resultant impact on the speed and scope of AI cyber tools will likely have destabilizing effects, nonetheless.[53]

In cyberspace, it is generally unclear in the early stages of a cyber operation whether an adversary intends to collect intelligence or prepare for an offensive attack, which is more likely to be used early on in a conflict scenario.[54] To illustrate these dynamics:[55] State A uses an AI-enhanced cyber-surveillance software (e.g. APT tools) into State B's command and control networks, supplying predictive AI algorithms with data that could be used by A to clandestinely build a baseline against B to support a pre-emptive attack in the future. Were State B to discover this intrusion, it might doubt the non-offensive nature of A's intrusions, especially if these networks were dual-use and used to support both B's conventional and nuclear forces. The US DoD's 2018 Cyber Strategy describes a 'defend forward' doctrine designed to "disrupt or halt malicious cyber activity at its source, including *activity that falls below the level of armed conflict*" (emphasis added).[56] Intrusive probing activities in cyberspace intended by one side as defensive measures, could (especially during a crisis) be misinterpreted by the targeted state as a threat to its conventional or nuclear forces.

As demonstrated in chapter 4, Chinese analysts view the vulnerability of China's NC3 systems to cyber infiltrations – even if an attacker aimed to collect information about cyber threats to prevent a future attack – as a highly escalatory national security threat.[57] By contrast, Russian analysts tend to view Russia's NC3 networks as relatively isolated, and thus, insulated from cyber-attacks.[58] Irrespective of these differences, the uncertainty caused by the possibility – actual or otherwise – that AI-enhanced 'cyber guns' might be used (or threatened) to undermine the credibility and reliability of states' nuclear forces would be inherently destabilizing. As the historical record attests, rational actors have incentives to misrepresent what they know (i.e. lie and bluff) to improve the terms of any settlement that may emerge from the wartime bargaining process, the risk-reward payoff.[59] A modicum of uncertainty about the effectiveness of AI-augmented cyber capabilities during a crisis or conflict would reduce *both* sides' risk tolerance, increasing the incentive to strike pre-emptively as a hedging strategy.

A virtuous cycle that might flow from enhanced reassurances premised on comprehensive intelligence would require information symmetry (i.e. equal access to intelligence and analysis systems) between states, and shared confidence in the accuracy and credibility of these systems. Perhaps most challenging of all in a world of 'revisionist' rising powers, the intentions of *all* states would need to be genuinely benign for these optimistic dynamics to play out[60] Moreover, like any form of intelligence, data gathered in cyberspace may be shared between militaries (especially among allies) to support missions that the originating state has no direct control over, with often troubling outcomes.[61]

During crisis conditions, for instance, an offensive AI cyber tool that succeeds in compromising an adversary's nuclear weapon systems, resulting in an 'asymmetric information' situation, could cause either or both sides to overstate (or understate) its retaliatory capabilities, and thus be more inclined to act in a risky and escalatory fashion.[62] In short, in a competitive strategic environment where states are inclined to assume the worst of others' intentions (described in chapters 3 and 4) one state's efforts to enhance the survivability of its strategic forces may be viewed by others as a threat to their nuclear retaliatory capability – or second-strike capacity.[63]

## Digital manipulation between nuclear states: rise of weaponized deepfakes?

AI ML techniques might also exacerbate the escalation risks by manipulating the digital information landscape, where decisions about nuclear weapons are made. In the competitive strategic environment, it is easy to imagine unprovoked escalation caused by malicious third-party (or state-proxy) clandestine false-flag operations.[64] During a crisis, the inability of a state to determine an attacker's intent may lead an actor to conclude that an attack – threatened or actual – was intended to undermine its nuclear deterrent.[65] For example, to incite conflict between two nuclear-armed rival states, State A uses proxy hackers to launch a deepfake video depicting senior military commanders of State B conspiring to launch a pre-emptive strike on State C.[66] Then, this footage is deliberately leaked into C's AI-augmented intelligence collection and analysis systems, provoking C to escalate the situation with strategic consequences. B, fearful of a decapitating strike and losing the first mover's advantage, swiftly escalates the situation.[67]

The current underdeveloped state of 'counter-AI' capabilities and other fail-safe mechanisms (e.g. circuit breakers) to de-escalate escalation in cyberspace will make the unprovoked and unintentional escalation dynamics depicted in this scenario very challenging to anticipate and, thus, mitigate.⁶⁸ As chapter 7 demonstrates, AI systems will likely come under increasing stress from nefarious attacks using counter-AI techniques (e.g. data pollution, spoofing, false alarms, or tricking a system in order to reverse engineering algorithms), which might undermine the confidence in a network, creating new vulnerabilities, errors, and unintentional escalation risks. Moreover, in the emerging deepfakes arms race (much like cybersecurity more generally), detection software will likely lag behind advances in offensive enabling solutions – or offense-dominant ones. According to computer science expert, Hany Farid, there are probably 100 to 1,000 times "more people developing the technology to manipulate content than there is to detect [it]."⁶⁹

Human cognition, and thus, effective deterrence and signaling intentions, is predicated on reliable and clear information; if an adversary is concerned that the information available to them is limited (or worse, inaccurate), they will likely assume the worst and act accordingly. Asymmetric information situations between rivals about the balance of military power could threaten crisis stability and, in turn, create rational incentives to escalate to nuclear confrontation. Consequently, states will be more inclined to assume the worst of others' intentions, particularly in situations where the legitimacy of the status quo is contested (e.g. maritime Asia).

As chapter 8 will illuminate, overreliance on automation (or 'automation bias') in the deployment of increasingly complex AI-enhanced applications – such as cyber, C3I systems, autonomous weapon systems, and precision missile munitions, designed to provide a distinct tactical advantage of machine-speed – will increase the vulnerability of these capabilities to exploitation. Moreover, the increasing substitution of human cognition for logic derived from machines will likely increase the opportunities for adversaries to exploit the limitations of 'narrow' AI technology (i.e. lack of human intuition, brittleness in complex real-world situations, and the inability to detect and counter manipulation attacks effectively).⁷⁰

## Conclusion

This chapter examined how AI-infused cyber capabilities might be used to subvert, or otherwise compromise, the reliability, control, and use of states' nuclear forces. The case study's key findings can be summarized as follows. First, AI applications designed to enhance cybersecurity for nuclear forces might simultaneously make cyber-dependent nuclear support systems (i.e. communications, data processing, or early warning sensors) more susceptible to cyber-attacks. A shared concern by great military powers about the potential threats posed by AI-augmented cyber warfare might impel states to adopt – or become more reliant on – a launch-on-warning nuclear posture.[71]

Second, and paradoxically, new technologies designed to enhance information can also undermine clear and reliable information flow and communication, critical for effective deterrence. Technologies that improve the reliability and speed of information processed might also increase the vulnerabilities of these networks. During crisis conditions, these vulnerabilities could create new first-mover advantages and increase escalation risks.

Third, by enabling improvements to the cyber offense, advances in AI might also exacerbate the cybersecurity challenge. Operating at machine speed, therefore, AI-augmented cyber tools could enable an attacker to exploit a narrow window of opportunity to penetrate an adversary's cyber-defenses or use APT tools to find new vulnerabilities faster and more easily than before.

Fourth, and related, AI-augmented cyber capabilities may create new pathways for accidental or inadvertent escalation. For example, a spoofing attack on an automated target recognition system's algorithm might escalate a situation based on a false premise that would unlikely fool the human eye. Moreover, the explainability – or the 'black box' – problem-set associated with AI could compound these escalation dynamics.

Fifth, AI ML techniques may also exacerbate the escalation risks by manipulating the digital information landscape, where decisions about the use of nuclear weapons are made. In competitive strategic environments, militaries could respond to an unprovoked escalation by a malicious third-party (or state-proxy), for example, by putting their nuclear weapons on high alert status. Asymmetric information situations between rivals about the balance of military power could threaten crisis stability and, in turn, create rational incentives to escalate a situation.

Finally, the current under-developed state of 'counter-AI' capabilities (and other fail-safe mechanisms) to de-escalate a situation will likely exacerbate the escalation risks associated with AI-augmented cyber capabilities. Therefore, until researchers unravel some of the unexplainable features of AI, human error and machine error will likely compound one-another, with unpredictable results.[72] In short, we are at a critical crossroads in the parallel (and symbiotic) evolution of the AI and cyberspace that national security communities globally will need to prepare for proactively.[73] In what direction could AI and autonomy be taking command and control systems and early warning systems? How vulnerable are modern C3I networks to the cyber offense? The next chapter investigates the role of AI in the strategic decision-making process.

## Notes

1. This chapter is derived in part from an article published in *Journal of Cyber Policy*, December 9, 2019, copyright Taylor & Francis, available online: https://doi.org/10.1080/23738871.2019.1701693 (accessed March 10, 2020).
2. Most defense analysts agree that cyber-warfare is 'offensive-dominant' in nature. See Erik Gartzke and Jon R. Lindsay, "Weaving Tangled Webs: Offense, Defense, and Deception in Cyberspace," *Security Studies*, 24, 2 (2015), pp. 316–348; Wyatt Hoffman, "Is Cyber Strategy Possible?" *The Washington Quarterly*, 42, 1 (2019), pp. 131–152; and Rebecca Slayton, "What is Cyber Offense–Defense Balance?" *International Security*, 41, 3 (2017), pp. 72–109. For an opposing view see Thomas Rid, "Think Again: Cyberwar," *Foreign Policy*, March/April 2012, pp. 80–84.
3. Daniel R. Coats, *Worldwide Threat Assessment of the US Intelligence Community*, January 29, 2019, www.dni.gov/files/ODNI/documents/2019-ATA-SFR-SSCI.pdf (accessed September 10, 2019).
4. The offense–defense balance is derived from a broader security dilemma theory that posits states may find themselves in conflict through their attempts to increase security (i.e. acquiring weapons and offensive strategies), inadvertently making another state feel less secure. See Charles Glaser and Chairn Kaufmann, "What is the Offense–Defense Balance and How Can We Measure It?" *International Security*, 22, 4 (1998), pp. 44–82; and Sean Lynn-Jones, "Offense–Defense Theory and Its Critics," *Security Studies*, 4, 4 (1995), pp. 660–691.
5. Bernard Brodie, *Strategy in the Missile Age* (Princeton, NJ: Princeton University Press, 1959), p. 175.
6. 'Deep learning' is a type of representation learning, which in turn is a type of machine learning. This technique can enhance the ability of machines to extract salient features from a landscape or image, which could be used for classification and pattern recognition. AI-augmented object and pattern recognition software supports military applications, including the DoD's Project Maven. In recent years, deep learning has become the most fashionable approach to AI engineering, but this does not mean it has supplanted other approaches. Ian Goodfellow, Yoshua Bengio, and Aaron Courville, *Deep Learning* (Cambridge, MA: MIT Press, 2016), p. 8.
7. US DARPA, the Intelligence Advanced Research Projects Activity (IARPA), and the National Science Foundation have all funded multiple projects involving the use of AI to enhance cyber defense. In 2016, for example, US (IARPA) launched its cyber-attack 'Automated Unconventional Sensor Environment program,' which seeks to develop innovative and automated methods to detect and predict cyberattacks appreciably earlier than existing approaches. Executive Office of the President of the US, "2016–2019 Progress Report on Advancing Artificial Intelligence R&D," November 2019, www.whitehouse.gov/wp-content/

uploads/2019/11/AI-Research-and-Development-Progress-Report-2016-2019. pdf (accessed September 10, 2019).
8 Research in the field of 'counter-AI' is still at a very nascent stage of development. Recently, one expert estimated that only around 1% of AI research spending today is allocated to security. Benjamin Buchanan, "The Future of AI and Cybersecurity," *The Cipher Brief*, October 30, 2019, www.thecipherbrief.com/column_article/the-future-of-ai-and-cybersecurity (accessed March 10, 2019).
9 US Department of Defence, "Summary of the 2018 Department of Defense Artificial Intelligence Strategy," February 2019, https://media.defense.gov/2019/Feb/12/2002088963/-1/-1/1/SUMMARY-OF-DOD-AI-STRATEGY.PDF (accessed September 10, 2019).
10 Savia Lobo, "The US DoD wants to dominate Russia and China in Artificial Intelligence. Last week gave us a glimpse into that vision," *Packt*, March 18, 2019, https://hub.packtpub.com/the-u-s-dod-wants-to-dominate-russia-and-china-in-artificial-intelligence-last-week-gave-us-a-glimpse-into-that-vision/ (accessed September 10, 2019).
11 See David F. Rudgers, "The Origins of Covert Action," *Journal of Contemporary History*, 35, 2 (April 2000), pp. 249–262.
12 Noah Shachtman, "Exclusive: Computer Virus Hits US Drone Fleet," *Wired*, October 7, 2011, www.wired.com/2011/10/virus-hits-drone-fleet/ (accessed September 10, 2019).
13 Tim Maurer, *Cyber Mercenaries: The State, Hackers, and Power* (Cambridge: Cambridge University Press, 2017), pp. 53–54.
14 Robert Jervis, "Some Thoughts on Deterrence in the Cyber Era," *Journal of Information Warfare* 15, 2 (2016), pp. 66–73.
15 Unlike nuclear deterrence, cyber deterrence does not necessarily fail if an attack is launched. Wyatt Hoffman, "Is Cyber Strategy Possible?" *The Washington Quarterly*, 42, 1 (2019), p. 143.
16 Several scholars argue that the use of offensive cyber capabilities at a strategic level of conflict is either operationally challenging or subject to successful deterrence from adversaries. Consequently, the cyber offense is considered most effective in 'grey zone' or low-level intensity use of force. See Slayton, "What is Cyber Offense–Defense Balance?" pp. 72–109; Jon R. Lindsay and Erik Gartzke, "Thermonuclear cyberwar," *Journal of Cybersecurity*, 3, 1 (2017), pp. 37–48; and Marti Libicki, *Cyberspace in Peace and War* (Annapolis: Naval Institute Press, 2016).
17 A counterargument posits that fragmented and diffused information vulnerabilities associated with cyberspace and digital networks more broadly are less likely to cause destabilizing first-strike incentives. See Thomas Rid, *Cyber War Will Not Take Place* (New York: Oxford University Press, 2013).
18 James S. Johnson, "The AI-Cyber Nexus: Implications for Military Escalation, Deterrence, and Strategic Stability," *Journal of Cyber Policy*, 4, 3 (2019), pp. 442–460.
19 Patricia Lewis and Beyza Unal, *Cybersecurity of Nuclear Weapons Systems:*

*Threats, Vulnerabilities and Consequences* (London: Chatham House Report, Royal Institute of International Affairs, 2018).
20 Recent reports of successful cyber-attacks against dual-use early warning systems suggest these claims are credible.
21 China and Russia have already taken steps towards placing their retaliatory forces on higher alert, and the fear of a US cyberattack against their nuclear-deterrent forces could prompt them to accelerate efforts to place more of their nuclear weapons on hair-trigger alert. See Gregory Kulacki, "China's Military Calls for Putting Its Nuclear Forces on Alert," *Union of Concerned Scientists*, January 2016, www.ucsusa.org/sites/default/files/attach/2016/02/China-Hair-Trigger-full-report.pdf (accessed September 10, 2019).
22 In recognition of this potential threat, a draft version of the US Congress National Defense Authorization Act for FY2020 required the Pentagon to develop a plan for ensuring the resiliency of nuclear C3I systems, including the possibility of negotiating a ban on cyber and other attacks against US adversary's (i.e. Chinese and Russian) networks, to avoid miscalculation and inadvertent nuclear war. Theresa Hitchens, "HASC Adds NC3 Funds; Wants Talks with Russia and China," *Breaking Defense*, June 10, 2019, https://breakingdefense.com/2019/06/hasc-adds-nc3-funds-wants-talkswith-russia-china/ (accessed March 10, 2020).
23 The cost of tools used to create malicious documents depends a lot on whether the malware can remain within the systems and escape detection by antivirus software. "Hack at All Cost: Putting a Price on APT attacks," August 22, 2019, www.ptsecurity.com/ww-en/analytics/advanced-persistent-threat-apt-attack-cost-report/ (accessed September 10, 2019).
24 Joseph F. Dunford, speech quoted at Jim Garamone, "Dunford: Speed of Military Decision-Making Must Exceed Speed of War," US Department of Defense, January 31, 2017, www.jcs.mil/Media/News/News-Display/Article/1067479/dunford-speed-of-military-decision-making-must-exceed-of-war/ (accessed September 10, 2019).
25 Paul Ingram, "Hacking UK Trident: A Growing Threat," *BASIC*, May 31, 2017, www.basicint.org/wp-content/uploads/2018/06/HACKING_UK_TRIDENT.pdf (accessed September 10, 2019).
26 AI machine learning systems rely on high-quality data-sets to train their algorithms, thus, injecting so-called 'poisoned' data into those training sets could lead these systems to perform in undesired and potentially undetectable ways.
27 Eugene Bagdasaryan et al., "How to Backdoor Federated Learning," *arXiv*, August 6, 2018, https://arxiv.org/pdf/1807.00459.pdf (accessed February 8, 2021).
28 'The internet of things' is the interconnectivity between physical objects, such as a smartphone or electronic appliance, via the internet, which allows these objects to collect and share data.
29 A recent survey reported that 70 percent of IoT devices lack rudimentary security safeguards. Michael Chui, Markus Löffler, and Roger Roberts, "The Internet

30 James R. Clapper, Director of National Intelligence, "Statement for the Record Worldwide Threat Assessment of the US Intelligence Community Senate Armed Services Committee," *US Office of the Director of National Intelligence*, February 6, 2016, www.armed-services.senate.gov/imo/media/doc/Clapper_02-09-16.pdf (accessed December 10, 2019).
31 During the 2003 invasion of Iraq, for example, a Patriot missile that shot down a British jet, killing both crewmen, was caused by the failure of humans in the loop to override an erroneous automated decision to fire.
32 James Johnson and Eleanor Krabill, "AI, Cyberspace, and Nuclear Weapons," *War on the Rocks*, January 31, 2020, https://warontherocks.com/2020/01/ai-cyberspace-and-nuclear-weapons/ (accessed March 10, 2020).
33 In this fictional scenario, if State A deliberately hacks State B's systems to label a civilian target as military, the only plausible reason for State A to do this is to trigger escalation. Thus, escalation in this case would be deliberate, not unintentional on the part of the initiator (State A).
34 For analysis on this point, see Martin Libicki, "A hacker way of warfare," in Nicholas D. Wright (ed.), *AI, China, Russia, and the Global Order: Technological, Political, Global, and Creative Perspectives*, Strategic Multilayer Assessment Periodic Publication (Washington, DC: Department of Defense, December 2018), pp. 128–132.
35 AI experts have proven that even when data appears accurate to AI image recognition software, these systems often 'hallucinate' objects that do not exist. Anna Rohrbach et al., "Object Hallucination in Image Captioning," *ariv.org*, https://arxiv.org/abs/1809.02156 (accessed December 10, 2019).
36 This fictional illustration demonstrates one possible (and worst case) outcome of a malicious operation against AI-enhanced weapon systems in an adversarial and human–machine collaborative scenario. Alternative outcomes could be conceived whereby advances in detection algorithms and sensor technology enable a human operator to avert a crisis before military force is used. Other safety mechanisms (i.e. circuit breakers or redundancies build into the systems) might also prevent this situation from spiraling.
37 Stephen J. Cimbala, *Nuclear Weapons in the Information Age* (New York: Continuum International Publishing, 2012); and Andrew Futter, *Hacking the Bomb: Cyber Threats and Nuclear Weapons* (Washington, DC: Georgetown University Press, 2018).
38 'Explainability' is a term used by AI experts referring to the fact that many AI systems produce outcomes with no explanation of the path the system took to derive the solution.
39 For example, it is uncertain how independently integrated AI platforms might interact with one another in a cross-domain battlefield environment. Technology for Global Security and Center for Global Security Research, "AI and the Military: Forever Altering Strategic Stability," *Technology for Global Security*,

www.tech4gs.org/ai-and-the-military-forever-altering-strategic-stability.html, p. 12 (accessed December 10, 2019).
40 Daniel S. Hoadley and Nathan J. Lucas, *Artificial Intelligence and National Security* (Washington, DC: Congressional Research Service, 2017), https://fas.org/sgp/crs/natsec/R45178.pdf, p. 2 (accessed August 10, 2019).
41 This problem-set is one of the main reasons why the use of AI ML in the context of weapon systems has, thus far, been limited to experimental research. Heather Roff and P.W. Singer, "The next president, will decide the fate of killer robots and the future of war," *Wired*, September 6, 2016, www.wired.com/2016/09/next-president-will-decide-fate-killer-robots-future-war/ (accessed December 10, 2019).
42 This paradox suggests that when states' capabilities are dependent resources (e.g. manpower or data-sets) that can be exploited or controlled by an adversary, both sides have incentives for a first strike. Jacquelyn Schneider, "The Capability/Vulnerability Paradox and Military Revolutions: Implications for Computing, Cyber, and the Onset of War," *Journal of Strategic Studies*, 42, 6 (2019), pp. 841–863.
43 For example, heightened escalation risks caused by aggressive US–Soviet expansion of counterforce technology during the Cold War reflected shifting nuclear doctrines on both sides (i.e. away from assured mutual destruction), not the pursuit of these technologies themselves. Austin Long and Brendan Rittenhouse Green, "Stalking the Secure Second Strike: Intelligence, Counterforce, and Nuclear Strategy," *Journal of Strategic Studies*, 38, 1–2 (2014), pp. 38–73.
44 Kris Osborn, "Navy Cyber War Breakthrough – AI Finds Malware in Encrypted Traffic," *Warrior Maven*, November 30, 2018, https://defensemaven.io/warriormaven/cyber/navy-cyber-warbreakthrough-ai-finds-malware-in-encrypted-traffic-K_tLobkkJkqadxDT9wPtaw/ (accessed January 5, 2019).
45 Libicki, *Cyberspace in Peace and War*.
46 A counterargument posits that fragmented and diffused information vulnerabilities associated with cyberspace and digital networks more broadly are less likely to cause destabilizing incentives like a first strike. See Rid, *Cyber War Will Not Take Place*.
47 This vulnerability is particularly concerning, given the dual-use nature of AI technology.
48 Paul Ingram, "Hacking UK Trident: A Growing Threat," *BASIC*, May 31, 2017, https://basicint.org/publications/stanislav-abaimov-paul-ingram-executive-director/2017/hacking-uk-trident-growing-threat (accessed December 10, 2019).
49 It may be difficult for a state to signal its intentions (i.e. offensive or defensive) in cyberspace to adversaries without subjecting the capability to relatively easy countermeasures (e.g. software patches or updates). See Austin Long, "A Cyber SIOP? Operational considerations for strategic offensive cyber planning," in Herbert Lin and Amy Zegart (eds), *Bytes, Bombs, and Spies: The Strategic Dimensions of Offensive Cyber Operations* (Washington, DC: Brookings Institution Press, 2019), pp. 105–132.

50 We saw this vulnerability all too clearly when a hacker brought a Jeep to a standstill on a busy highway and interfered with its steering system, accelerating it. Andy Greenberg, "The Jeep Hackers are Back to Prove Car Hacking Can Get Much Worse," *Wired*, August 1, 2016, www.wired.com/2016/08/jeep-hackers-return-high-speed-steering-acceleration-hacks/ (accessed December 10, 2019).
51 A 'left-of-launch operation' refers to a cyber-offensive operation that would defeat the threat of a nuclear ballistic missile before it is launched. Recently, there has been discussion in the US on the possibility of using 'left of launch' cyberattacks against North Korea's nuclear forces. That is, signaling to the North Korean leadership the ability of the US to use these capabilities to nullify the threat of a nuclear strike. See William Broad and David Sanger, "US Strategy to Hobble North Korea Was Hidden in Plain Sight," *New York Times*, March 4, 2017, www.nytimes.com/2017/03/04/world/asia/left-of-launch-missile-defense.html (accessed December 10, 2019).
52 For more on this topic, see Ben Buchanan, *The Cybersecurity Dilemma* (New York: Oxford University Press, 2017); and Ben Buchanan and Taylor Miller, "Machine Learning for Policymakers," Paper, *Cyber Security Project*, Belfer Center, 2017, www.belfercenter.org/sites/default/files/files/publication/MachineLearningforPolicymakers.pdf (accessed December 10, 2019).
53 Many AI additions that can augment offensive cyber capabilities involve either enumerating the target space or repackaging malware to avoid detection.
54 Herbert Lin, "Reflections on the New Department of Defense Cyber Strategy: What It Says, What It Doesn't Say," *Georgetown Journal of International Affairs*, 17, 3 (2016), pp. 5–13.
55 Johnson and Krabill, "AI, Cyberspace, and Nuclear Weapons."
56 DoD, "Summary: Department of Defense Cyber Strategy 2018," September 2018, https://media.defense.gov/2018/Sep/18/2002041658/-1/1/1/CYBER_STRATEGY_SUMMARY_FINAL.PDF (accessed December 10, 2019).
57 James M. Acton (ed.), with Li Bin, Alexey Arbatov, Petr Topychkanov, and Zhao Tong, *Entanglement: Russian and Chinese Perspectives on Non-Nuclear Weapons and Nuclear Risks* (Washington, DC: Carnegie Endowment for International Peace, 2017), p. 81.
58 Ibid.
59 James D. Fearon, "Rationalist Explanations for War," *International Organization*, 49, 3 (1995), pp. 379–414.
60 The White House, *National Security Strategy of the United States of America*, December 2017, www.whitehouse.gov/wp-content/uploads/2017/12/NSS-Final-12-18-2017-0905.pdf (accessed December 10, 2019).
61 For example, in 2012 – under the terms of an intelligence-sharing agreement signed in 2007 – the US military shared the targeting information from a Predator drone to the Turkish military, which launched an airstrike on the convoy that killed 34 civilians. Adam Entous and Joe Parkinson, "Turkey's Attack on Civilians Tied to US Military Drone," *The Wall Street Journal*, May 16, 2012, www.wsj.com/articles/SB10001424052702303877604577380480677575646 (accessed December 10, 2019).

62 The literature about deception and capabilities demonstrates that states often have political incentives to conceal, lie about, or exaggerate their military strength, rather than reveal or use it for signaling purposes. See Barton Whaley, "Covert Rearmament in Germany, 1919–1939: Deception and Misperception," *Journal of Strategic Studies*, 5, 1 (March 1982), pp. 3–39; John J. Mearsheimer, *Why Leaders Lie: The Truth about Lying in International Politics* (Oxford: Oxford University Press, 2013); and Robert L. Jervis, *The Logic of Images in International Relations* (New York: Columbia University Press, 1989).

63 Thomas C. Schelling and Morton Halperin, *Strategy and Arms Control* (New York: The Twentieth Century Fund 1961), p. 30.

64 See Martin C. Libicki, *Cyber Deterrence and Cyberwar* (Santa Monica, CA: RAND, 2009), p. 44.

65 For example, even if the malware detected in an attack was only capable of espionage, a target may fear that it also contained a 'kill switch' able to disable an early warning system after activation.

66 Generative adversarial networks (GANs) are a new approach, which involves two artificial neural network systems that spar with each other to create a realistic original image, audio, or video content, which machines have never been able to do properly before. Karen Hao, "Inside the World of AI that Forges Beautiful Art and Terrifying Deepfakes," *MIT Technology Review*, January 4, 2019, www.technologyreview.com/s/612501/inside-the-world-of-ai-that-forges-beautiful-art-and-terrifying-deepfakes/ (accessed December 10, 2019).

67 Similar to the previous fictional case study, alternatives to this worst scenario outcome can also be conceived. At the most basic level, the rung of the escalation ladder that led to catastrophe (i.e. the source of the GANs), and subsequent subversive tactics, could have been detected and, thus, crisis averted. The case study also assumes the victims of this offensive operation viewed each other as adversaries, making them more suspicious of the other's intentions in the event of a crisis. Were *either* side able to demonstrate empathy or display restraint, this escalatory outcome may be avoided.

68 'Counter-AI' broadly includes security measures within AI systems to protect them from subversion and manipulation, as well as efforts to deter and defend against the malicious use of AI. While research in the field of 'counter-AI' is still at a very nascent stage, analysts have made some progress in detecting anomalies in network behavior as a means to isolate possible exploitable vulnerabilities within machine learning AI software. Scott Rosenberg, "Firewalls Don't Stop Hackers, AI Might," *Wired*, August 27, 2017, www.wired.com/story/firewalls-dont-stop-hackers-ai-might/ (accessed December 10, 2019).

69 Richard Fontaine and Kara Frederick, "The Autocrat's New Tool Kit," *The Wall Street Journal*, March 15, 2019, www.wsj.com/articles/the-autocrats-new-tool-kit-11552662637 (accessed December 10, 2019).

70 One potential countermeasure could be to train each AI system in a slightly different way, to limit the success of system failures caused by exploitation to result in the collapse of an entire overall military operation. Libicki, "A Hacker Way of Warfare," pp. 128–132.

71 In addition to cyber capabilities, other advanced weapon systems, such as hypersonic weapons (see chapter 6), could also prompt states to adopt launch-on-warning, or pre-emptive policies to reduce the vulnerability of their nuclear weapons.
72 Generally, the less predictable the environment, the harder it is to model and, thus, the more difficult it becomes to create autonomous capabilities within systems that are effective, safe, and reliable. See Kimberly A. Barchard and Larry A. Pace, "Preventing Human Error: The Impact of Data Entry Methods on Data Accuracy and Statistical Results," *Computers in Human Behavior* 27 (2011), pp. 1834–1839.
73 George Dvorsky, "Hackers Have Already Started to Weaponize Artificial Intelligence," *Gizmodo*, November 9, 2017, https://gizmodo.com/hackers-have-already-started-to-weaponize-artificial-in-1797688425 (accessed December 10, 2019).

# 8

# Delegating strategic decisions to intelligent machines

Will the use of AI in strategic decision-making be stabilizing or destabilizing? How might synthesizing AI with nuclear command, control, and communications (NC3) early warning systems impact the nuclear enterprise?[1] The compression of detection and decision-making timeframes associated with the computer revolution is not an entirely new phenomenon (see chapter 2). During the Cold War, the US and Soviet Union both automated their nuclear command-and-control, targeting, and early warning detection systems to strengthen their respective retaliatory capabilities against a first strike.[2] Technologies developed during the 1950s paved the way for modern undersea sensors, space-based communication, and over-the-horizon radars.[3] Moreover, many of the systems and concepts introduced in the 1960s are still in use today. For example, the first Defense Support Program satellite for ballistic missile launch warning, launched in 1970, remains a core element of US early warning infrastructure.[4]

Sophisticated NC3 networks interact with nuclear deterrence through several vectors:[5] (1) early warning satellites, sensors, and radars (e.g. to detect incoming missile launches); (2) gathering, aggregating, processing, and communicating intelligence for command and control (C2) planning (i.e. to send and receive secure and reliable orders and status reports between civilian and military leaders);[6] (3) missile defense systems as a critical component of nuclear deterrence and warfighting postures; and (4) monitoring, testing, and assessing the security and reliability of sensor technology, data, and communications channels, and weapon launch platforms, used in the context of NC3.[7] NC3 systems supply critical linkages between states' nuclear forces and their leadership, ensuring decision-makers have the requisite information and time needed to command and control nuclear forces.

In short, NC3 systems are a vital pillar of the states' deterrence and communications – to ensure robust and reliable command and control over nuclear weapons under all conditions – and can have a significant impact on how wars are fought, managed, and terminated. Some analysts refer to NC3 as the 'fifth pillar' of the nuclear deterrence; after the three legs of the triad and nuclear weapons themselves.[8] Because of the pivotal nature of these systems to the nuclear enterprise, superior systems would likely outweigh asymmetries in arsenals sizes – and thus put an adversary with less capable systems and more missiles at a disadvantage.

Nuclear security experts have cataloged a long list of computer errors, faulty components, early warning radar faults, lack of knowledge about adversaries' capabilities and *modus operandi* (especially missile defense systems), and human mistakes that led to nuclear accidents and demonstrated the limitations of inherently vulnerable NC3 systems and the potential for malicious interference.[9] The risks and trade-offs inherent in NC3 systems since the Cold War-era, reflecting the complex interactions of tightly coupled systems and social, emotional, heuristic, and cognitive evolution of human agents, making decisions amid uncertainty, will likely be amplified by the inexorable and ubiquitous complexity and unpredictability that AI introduces.[10]

AI-augmented systems operating at machine speed, and reacting to situations in ways that may surpass humans' comprehension, would challenge the natural strategic-psychological abilities of commanders (i.e. intuition, flexibility, heuristics, and empathy) in decision-making, raising broader ethical questions about escalation control and the start of a slippery slope towards the abandonment of human moral responsibility.[11] That is, the uncertainties and unintended outcomes of machines interpreting human intentions, and making autonomous strategic decisions, in fundamentally non-human ways.

The remainder of this chapter proceeds in two parts. The first describes how defense planners might use AI in the strategic decision-making process. It examines the notion of human psychology to elucidate how and why military commanders might use AI in this process, despite the concerns of defense planners. It considers the risks and trade-offs of increasing – inadvertently or otherwise – the role of machines in the strategic decision-making process. How might variations such as regime type, nuclear doctrine, strategy, strategic culture, or force structure make states more (or less) predisposed to use AI in strategic decision-making? The second part of the chapter considers the implications for the nuclear enterprise of synthesizing AI with NC3 early warning systems. Will this synthesis be stabilizing or destabilizing? How might non-nuclear states (and non-state actors) leverage AI to place pressure on nuclear states?

## AI strategic decision-making: magic 8-ball?

The recent success of AI systems in several highly complex strategic games has demonstrated insightful traits that have potentially significant implications for future military-strategic decision-making.[12] In 2016, for example, DeepMind's AlphaGo system defeated the professional Go master, Lee Sedol. In one game, the AI player reportedly surprised Sedol in a strategic move that "no human would ever do."[13] Three years later, DeepMind's AlphaStar system defeated one of the world's leading e-sports gamers at StarCraft II – a complex multiplayer game that takes place in real-time and in a vast action space with multiple interacting entities – and devised and executed complex strategies in ways a human player would unlikely do.[14] In short, existing rule-based machine learning (ML) algorithms would likely be sufficient to automate C2 processes further.

The central fear of alarmists focuses on two related concerns. First, the potential dystopian (e.g. the *Terminator's* Skynet prophetic imagery) and existential consequences of AI surpassing human intelligence. Second, the possible dangers caused by machines that, absent of human empathy (or other 'theory of the mind' emotional attributes),[15] relentlessly optimize pre-set goals – or self-motivated future iterations that pursue their own – with unexpected and unintentional outcomes – or *Dr. Strangelove* doomsday machine comparisons.[16] Without this natural ability, how might machines (mis)interpret human signals of resolve and deterrence, for example?

Human commanders supported by AI – functioning at higher speeds, using non-human processes, and compressed decision-making timeframes in complex situations with incomplete information – might impede the ability (or Clausewitzian 'genius') of commanders to shape the actions of AI-augmented autonomous weapon systems.[17] For now, there is general agreement among nuclear-armed states that even if technological developments allow, decision-making that directly impacts the nuclear command and control should not be pre-delegated to machines – not least because of the explainability, transparency, and unpredictability problems associated with ML algorithms.[18]

Psychologists have demonstrated that humans are slow to trust the information derived from algorithms (e.g. radar data and facial recognition software), but as the reliability of the information improves so the propensity to trust machines increases – even in cases where evidence emerges that suggests a machine's judgment is incorrect.[19] The tendency of humans to use automation (i.e. automated decision support aids) as a heuristic replacement for vigilant information seeking, cross-checking, and adequate processing supervision, is known as 'automation bias.' Despite humans' inherent distrust of machine-generated information, once AI demonstrates an apparent capacity to engage with, and interact in, complex military situations (i.e. wargaming) at a human (or superhuman) level, defense planners will likely become more predisposed to view decisions generated by AI algorithms as analogous (or even superior) with those of humans – even if these decisions lack sufficiently compelling 'human' rational or fuzzy 'machine' logic.[20]

Experts have long recognized the epistemological and metaphysical confusion that can arise from mistakenly conflating human and machine intelligence, used, in particular, in safety-critical high-risk domains.[21] Human psychology research has found that people are predisposed to harm others if ordered to do so by an authority figure.[22] As AI-enabled decision-makings tools are introduced into militaries, human operators may begin to view these systems as agents of authority (i.e. more intelligent and more authoritative than humans), and thus be more inclined to follow their recommendations blindly; even in the face of information that indicates they would be wiser not to.

This predisposition will likely be influenced, and possibly expedited by human bias, cognitive weaknesses (notably decision-making heuristics), assumptions, and the innate anthropomorphic tendencies of human psychology. Studies have shown that humans are predisposed to treat machines (i.e. automated decision support aids) that share task-orientated responsibilities as 'team members,' and in many cases exhibit similar in-group favoritism as humans do with one another.[23] The delegation of responsibility to machines might dilute the sense of shared responsibility in this human–machine collaboration.[24] In turn, this could cause a reduction in human effort and vigilance and increase the risk of errors and accidents.[25]

Contrary to conventional wisdom, having a human in the loop in decision-making tasks also appears not to alleviate automation bias.[26] Instead, and surprisingly, human–machine collaboration in monitoring and sharing responsibility for decision-making can lead to similar psychological effects that occur when humans share responsibilities with other humans, whereby 'social loafing' arises – the tendency of humans to seek ways to reduce their effort when working redundantly within a group compared to when they work individually on a task.[27] A reduction in human effort and vigilance caused by these tendencies could increase the risk of unforced error and accidents. Besides, a reliance on the decisions of automation in complex and high-intensity situations can make humans less attentive to – or more likely to dismiss – contradictory information, and more predisposed to use automation as a heuristic replacement (or shortcut) for information seeking.[28]

The decision to automate nuclear assets might also be influenced by the political stability and the threat perceptions of a nuclear-armed state. A regime that fears either a domestic-political challenge to its rule or foreign interference may elect to automate its nuclear forces, ensuring only a small number of agents are involved in the nuclear enterprise.[29] China, for example, maintains strict controls on its nuclear command and control structures (i.e. separating nuclear warhead and delivery systems). Open-source evidence does not suggest Beijing has pre-delegated launch authority down the chain of command if a first strike decapitates the leadership. In short, as a means to retain centralized command and control structures and strict supervision over the use of nuclear weapons, AI-enabled automation might become an increasingly amenable option to authoritarian regimes such as China.

Authoritarian states may perceive an adversary's intentions differently from a democratic one. The belief that a regime's political survival or legitimacy is at risk might cause leaders to consider worst-case scenario judgments and behave in a manner predicted by offensive realist scholars.[30] Conversely, non-democratic leaders operating in closed political systems such as China may exhibit a higher degree of confidence or sanguinity in their ability to respond to perceived threats in international relations. Bias assessments from a non-democratic regime's intelligence services might reinforce a leader's faith – or a false sense of security – in their diplomatic skill and maneuverability.[31]

Moreover, a regime that views its second-strike capabilities – including its NC3 systems – as vulnerable or insecure (e.g. North Korea or perhaps China) may be more inclined to automate its nuclear forces and launch postures. In short, non-democratic nuclear states with relatively centralized command and control structures, less confident in the survivability of their nuclear arsenal, and whose political legitimacy and regime stability is conditioned by the general acceptance of official narratives and dogma, would likely be more persuaded by the merits of automation, and less concerned about the potential risks – least of all the ethical, human cognitive, or moral challenges – associated with this decision. Despite official Chinese statements supporting the regulation of military AI by global militaries, much of China's AI-related initiatives (e.g. the use of data for social surveillance to distill a social-credit scoring system, and ubiquitous facial recognition policies) focus on the impact on social stability and, in particular, efforts to insulate the legitimacy of the regime against potential internal threats.[32]

By contrast, the political processes, accountability (especially elected leaders and head of state vis-à-vis public opinion), nuclear-launch protocols, nuclear strategy and doctrine, mature civil–military relations, and shared values between allies (i.e. US and its NATO allies), in democratic societies should make them less predisposed – or at least more reticent and encumbered – in use of AI in the nuclear domain.[33] Perhaps the question to ask is less *whether* AI will be integrated into NC3 systems, but rather by whom, to what extent, and at what cost to the nuclear enterprise.

## A prediction revolution and automated escalation

At a theoretical level, the Defense Advanced Research Projects Agency (DARPA)'s Knowledge-Directed Artificial Intelligence Reasoning Over Schemas (KAIROS) program demonstrates how NC3 systems infused with AI technology might function. KAIROS integrates contextual and temporal events of a nuclear attack into an analytics-based AI application, which can generate associated and prompt actionable responses.[34] KAIROS also highlights the need for AI-enhanced NC3 systems to handle the increasingly blurred lines between nuclear and conventional dual-use weapons and support systems. This blurring (or co-mingling) problem-set – described in chapters 2 and 4 – increases the potential for miscalculation and accidental escalation. As a corollary, AI-enhanced early warning detection systems must be able to reliably determine whether an imminent attack on its dual-use C3I systems (e.g. communication and surveillance satellites), during a conventional conflict, is intended as a non-nuclear offensive campaign, or as a prelude to escalation to nuclear confrontation.[35]

The biggest technical challenge for deploying a system like KAIROS is developing the ability to learn and adapt without the requirement for an iterative learning process – common in today's narrow AI applications such as Google Assistant, Google Translate language processing tools, and Google's AlphaGo supercomputer. Further, as chapter 1 shows, the algorithms that power AI systems like AlphaGo are usually trained on vast data-sets that are not available to support nuclear weapons. Designing and training an ML algorithm for a nuclear early warning would be almost entirely reliant on simulated data, which in the safety-critical nuclear domain would be extremely risky.[36]

Data limitations, coupled with constraints on the ability of AI algorithms to capture the nuanced, dynamic, subjective, and changeable nature of human commanders (or theory-of-the-mind functions) will mean that for the foreseeable future strategic decision-making will remain a fundamentally human endeavor – albeit imbued with increasing degrees of interaction with intelligent machines.[37] That is, AI will continue to include some human agency – especially in collaboration with machines – to effectively manage the attendant issues associated with technological complexity and independency, avoiding, for now, the risks associated with pre-delegating the use of military force.[38]

While human agency should ensure that the role of AI in the nuclear domain is confined to a predominately tactical one – through the discharge of its 'support role' – it might nonetheless (and possibly inadvertently) influence strategic decisions that involve nuclear weapons. The distinction between the impact of AI at a tactical and strategic level is not a binary one:[39] technology designed to augment autonomous tactical weapons ostensibly (e.g. NC3, ISR, and early warning systems) will make decisions in the use of lethal force, which informs and shapes strategic warfaring calculations.[40] The US 2018 *Nuclear Posture Review*, for example, explicitly states that the DoD would pursue design support technologies (such as ML) to facilitate more effective and faster strategic decision-making.[41] In short, escalation at the tactical level could easily have *strategic effects*.

To support officers' construct operational plans, DARPA has designed AI-powered supports systems (Integrated Battle Command and 'Deep Green') that allow commanders to visualize, evaluate, anticipate an adversary's strategic intentions, and predict the impact of complex environments with changing parameters.[42] Chinese analysts have also begun to research the use of big-data and deep-learning AI techniques to enhance the processing speed and intelligence analysis of satellite images and support the People's Liberation Army (PLA)'s early warning capabilities and enable a 'prediction revolution' in future warfare.[43]

In 2017, the PLA Rocket Force's Engineering University participated in an international workshop that was convened to focus on intelligent reasoning and decision-making.[44] Besides, China has also applied AI to wargaming and military simulations and researched AI-enabled data retrieval, and analysis from remote sensing satellites,[45] to generate data and insights that might be used to enhance Chinese early warning systems, situational awareness, and improve targeting.[46]

While AI-enabled decision support tools are not necessarily destabilizing, this non-binary distinction could risk AI 'support' tools substituting the role of critical thinking, empathy, creativity, and intuition of human commanders in the strategic decision-making process. The danger of delegating (inadvertently or otherwise) moral responsibility to machines raises broader issues about the degree of trust and reliance placed with these systems. Ethicists emphasize the need for AI as 'moral agents' to exhibit reasoning – as opposed to Bayesian reasoning and fuzzy logic – through not only careful reflection and deliberation but also an aptitude to effectively and reliably simulate human emotions (especially empathy) for interacting socially with humans in many contexts.[47]

Unlikely as it may be that commanders would explicitly delegate – at least knowingly – the authority of nuclear launch platforms (nuclear-powered ballistic missile submarines, bombers, missile launch facilities, and transporter erectors-launchers), or nuclear delivery vehicles and support systems (e.g. intercontinental ballistic missiles, torpedoes, missiles, nuclear-armed long-endurance unmanned aerial vehicles, and NC3) to machines, AI is expected to be more widely used to support decision-making on strategic issues.[48] If planners come to view this 'support' function as a panacea for the cognitive fallibilities of human analysis and decision-making, however, the reliance on these systems may have destabilizing consequences. As shown in chapter 1, AI ML algorithms are only as good as the data and information they are trained on and supplied with during operations. Further, because of the paucity of data available for AI to learn from in the nuclear domain, designing an AI-augmented support tool to provide early warning s systems with reliable information on a pre-emptive nuclear strike, therefore, would be extremely challenging.

Although AI systems can function at machine speed and precision in the execution of military force, algorithms adhering to pre-determined mission goals are unable to empathize with humans, which is necessary to determine or anticipate the intentions and behavior of an adversary – that is, intentions communicated in the use of military action to signal deterrence or resolve (i.e. the willingness to escalate) during a crisis. Machines would likely be worse (or at least less reliable) at understanding human signaling involved in deterrence, in particular signaling de-escalation.[49] Not only would machines need to understand human commanders and human adversaries, but they must also be able to interpret an adversary AI's signaling and behavior.

An AI algorithm that is optimized to pursue pre-programmed goals might, therefore, misinterpret an adversary is simultaneously signaling resolve while seeking to avoid conflict or de-escalate a situation. Without reliable means to attribute an actor's intentions, AI systems may convey undesirable and unintended (by human commanders) signals to the enemy, complicating the delicate balance between an actor's willingness to escalate a situation as a last resort and keeping the option open to step back from the brink.[50]

Counterintuitively, states may view increased NC3 automation as a means to manage escalation and enhance deterrence, signaling to an adversary that an attack (or the threat of one) could trigger a nuclear response – or the notion of 'rationality-of-irrationality.'[51] Because of the difficulty of demonstrating, and thus effectively signaling, this automation posture before a crisis or conflict, this implicit – and possibly unverifiable – threat would likely intensify crisis instability. In short, if a nuclear-armed state used automation to reduce its flexibility during a crisis, and without the ability to credibly signal this to an adversary, it would be akin to Thomas Schelling's game theory-inspired notion of tearing out the steering wheel in a game of chicken without being able to throw it out the window.[52] The uncertainty and strategic ambiguity caused by this posture could have a deterrent effect irrespective of whether this signal was credible or not.[53]

Furthermore, unwarranted confidence and reliance on machines – associated with automation bias – in the pre-delegation of the use of force during a crisis or conflict, let alone during nuclear brinkmanship, might inadvertently compromise states' ability to control escalation.[54] Overconfidence, caused or exacerbated by automation bias, in the ability of AI systems to predict escalation and gauge intentions – and deter and counter threats more broadly – could embolden a state (especially in asymmetric information situations) to contemplate belligerent or provocative behavior it might otherwise have thought too risky.[55]

This misplaced confidence might also reduce – or even eliminate – the psychological uncertainty that injects caution into defense planning, which may intensify existing escalation risks into a crisis or conflict.[56] For example, as shown in chapter 3, China's substantial investment and strategic interest in AI-augmented decision support systems are part of a broader doctrinal emphasis on the concept of information dominance through scientific central-planning – to enable the PLA to respond faster to a disarming attack – and suggests that Chinese commanders may be susceptible to automation bias.[57]

During nuclear brinkmanship, the ultimate competition in risk-taking, the interaction between machines and human-strategic psychology (or the war of ideas), caused by the pre-delegation of escalation to autonomous weapons (or automated escalation), may increase the risk of misinterpreting an adversary's intentions (or warning), thereby increasing the risks associated with closing the damage-limitation window, undermining crisis stability, and increasing first-strike incentives.[58] An ML algorithm that is programmed to optimize pre-programmed goals might misinterpret an adversary as simultaneously signaling resolve while seeking to avoid conflict or de-escalate a situation. Moreover, AI-controlled NC3 systems would be more vulnerable to subversion from cyber-attacks, which could increase this risk of inadvertent escalation caused by human or machine miscalculation or error – even if humans are kept 'in the loop.'[59]

As demonstrated in Part II, competitive pressures may result in the implementation of AI applications (both offense and defense) *before* they are sufficiently tested, verified, or technically mature, making these systems more error-prone and susceptible to subversion – in particular a cyber-attack.[60] Thus, even a well-fortified and fully trained AI system might remain vulnerable to subversion that would be difficult to detect and even harder to attribute.[61] A clandestine cyber-attack that undermines effective information and communication flow would risk increasing the incentives for escalation.[62] Although system inputs and outputs can be observed, the speed and scale of ML mechanisms would make it difficult for operators to isolate, and thus explain a particular machine-generated prediction or decision.[63]

## A double-edged sword for stability

AI-augmented support systems, and the expanded use of automation in NC3 more broadly, could, in several ways, improve confidence in the existing nuclear enterprise,[64] making accidents caused by human error – especially false warnings – and the unauthorized use of nuclear weapons less likely, thereby enhancing strategic stability. First and foremost, ML algorithms and autonomous systems may be used to bolster NC3 defenses against both physical (e.g. kinetic attacks against C2 nodes) or cyber-threats (e.g. offensive cyber, jamming attacks, and electromagnetic pulses generated by a high-altitude nuclear burst effort).

Second, AI-augmented communications systems could improve information flow and situational awareness, allowing militaries to operate at scale in complex environments, particularly in situations with incomplete information. In this way, as described in chapter 1, technology like AI might expand the decision-making timeframe available to commanders during a crisis, a perspective that has been overlooked by global defense strategic communities.

Improvements in AI-enhanced situational awareness and prediction capabilities, by shifting the decision-making authority away from combat personnel to senior commanders, might reduce the uncertainties caused by the delegation of tactical decision-making to individual soldiers – bringing to bear combat experience, the lessons of history, political acumen, and moral courage – who are, in theory,[65] best placed to determine tactical solutions, and entrusted to implement commanders' intent – or 'mission command.'[66] Whether these enhancements, by centralizing the decision-making process, and creating a new breed of 'tactical Generals' – micro-managing theater commanders from afar – will improve military effectiveness or exacerbate uncertainties is, however, an open question.[67]

Third, ML techniques coupled with advances in remote sensing technology might enhance nuclear early warning and testing systems, making accidents caused by error less likely. Finally, automating several NC3 functions may reduce the risk of human error caused by cognitive bias, repetitive tasks, and fatigue.[68] For instance, unmanned aerial vehicles might replace signal rockets to form an alternative airborne communications network, especially useful in situations where satellite communication is not possible. These enhancements could enable a variety of operations, inter alia, bolstering non-nuclear capabilities such as cyber, air defenses, and electronic jamming, improving target identification and pattern recognition systems, controlling autonomous platforms, and improving the way workforce and logistics are managed, to name but few.

DARPA's Real-Time Adversarial Intelligence and Decision-Making

(RAID) ML algorithm is designed to predict the goals, movements, and even the possible emotions of an adversary's forces five hours into the future. RAID relies on a type of game theory that shrinks down problems into smaller games, reducing the computational power required to solve them.[69] Similarly, BAE Systems is working with DARPA to design cognitive-based ML algorithms and data models aimed at giving space operators the ability to identify abnormal activities from vast data-sets, and to predict possible threats such as space-based launches and satellite movements.[70]

Notwithstanding the notable progress made in the use of ML techniques (notably neural networks) for conflict and escalation prediction, no system has currently emerged that can reliably capture the diverse variables, uncertainties, and complex interactions of modern warfare. Predicting when and where a conflict will break out, for example.[71] These predictive tools may, however, prove useful as heuristic decision-making tools to generate possible scenarios rather than producing actionable strategic policy advice.

As the technology matures, these systems may be able to identify risks and offer strategic options – including risks not anticipated by humans – and, ultimately, map out an entire campaign.[72] As a corollary, AI systems might react to dynamic and complex combat situations more rapidly, learn from their mistakes, and be burdened with fewer cognitive shortcomings than human commanders such as human emotion, heuristics, and group-think.

While state-of-the-art technology such as AI, ML, big-data analytics, sensing technology, quantum communications, and 5G-supported networks integrated with nuclear early warning systems, might alert commanders of incoming threats faster, the greater precision and scalability afforded by these advances could exacerbate escalation risks in two ways.[73] First, AI ML, which is used as a force multiplier for the cyber offense (e.g. data poisoning, spoofing, deepfakes, manipulation, hacking, and digital jamming), would be considerably more difficult for early warning systems to detect – or detect in time.[74] An adversary might, for example, target 'blind spots' in ML neural networks to nefariously manipulate data in such a way that both the human operator and AI would not recognize a change – known as data-poisoning or data-pollution.[75]

As demonstrated in chapter 7, an AI ML generative adversarial network (GAN) generated deepfake, and a data-poisoning attack could trigger an escalatory crisis between two or more nuclear states.[76] A state or non-state actor might, for example, use an image or recording of a military commander obtained from open sources to generate and disseminate a deepfake containing false orders, intelligence, or geospatial imagery that at best generates confusion, and at worst, aggravates a tense situation or crisis between rival nuclear powers – this theme is examined below.[77] In this sense, deepfakes will likely become (or already are) another capability in the toolkit of warfighters to wage campaigns of disinformation and deception – one that both sides may use of have used against them.[78]

Second, in the unlikely event that an attack, subversion, or manipulation was successfully detected, threat identification (or attribution) at machine speed would be virtually impossible. Once an operation is executed, human operators would be unable to monitor a system's decision calculus in realtime. Thus, the ability to effectively monitor and control escalation – or de-escalate a situation – would be impaired. Even if nuclear early warning systems eventually detected an intrusion, heightened levels of uncertainty and tension caused by an alert might impel the respective militaries to put their nuclear weapons on high alert status – to reduce the vulnerability of their strategic forces. In sum, asymmetric situations between adversaries could prompt states to shift their nuclear doctrine and postures (e.g. endorsing a doctrine of pre-emption or limited nuclear strikes), and expedite the reconciliation, an expanded use of AI in the nuclear domain – even at the expense of control and stability.[79]

During a crisis or conflict, the inability of a state to determine an attacker's intentions may lead it to conclude that an attack (threatened or actual) was designed to erode its nuclear deterrence – known as a 'false positive.'[80] Conversely, a malfunctioning early warning system caused by a malicious attack could mean a nuclear state is oblivious of an imminent nuclear attack, thus impeding it from responding appropriately due to degraded nuclear decision-making – known as a 'false negative.' For instance, China's fear that the PLA's early warning systems are inadequate to respond to the US disarming first nuclear strike prompted Chinese planners to prioritize a reduction of *false negatives* (misidentifying a nuclear weapon as a non-nuclear one) over *false positives* (misidentifying a non-nuclear weapon as a nuclear one) – emphasized by the US.[81] Both false positives and false negatives can cause misperceptions and mischaracterizations in ways that might exacerbate escalation risk. This skewed assessment in the context of nuclear weapons ready to launch at a moment's notice, could precipitate worst-case scenario thinking and trigger inadvertent escalation.[82]

According to open sources, operators at the North American Aerospace Defense Command (NORAD) have less than three minutes to assess and confirm initial indications from early warning systems of an incoming attack.[83] This compressed decision-making timeframe could put political leaders under intense pressure to decide whether to escalate during a crisis, with incomplete (and possibly false) information of a situation. In the context of advanced technologies, response systems like NORAD raise particular concerns: missiles cannot be recalled; submarines on deterrence patrols may be out of touch for extended times with a high degree of C2 autonomy; and nuclear weapons may be launched accidentally.[84] As

Thomas Schelling presciently warned during the Cold War, "when speed is critical, the victim of an accident or a false alarm is under terrible pressure."[85] Paradoxically, therefore, new technologies designed to enhance information (i.e. modernized NC3 systems supported by 5G networks, AI ML, big-data analytics, and quantum computing) might simultaneously erode precise and reliable information flow and communication, critical for effective deterrence.[86]

In addition to the nuclear interactions between nuclear-armed dyads, nefarious information manipulation by non-state actors (i.e. terrorists and criminals) and state proxy actors could have destabilizing implications on effective deterrence and military planning, both during times of peace and war.[87] AI-enhanced fake news, deepfakes, bots, and other malevolent social media campaigns could also influence public opinion – creating false narratives or amplifying false alarms – with destabilizing effects on a mass scale, especially in times of geopolitical tension and internal strife.[88] For example, in 2014, thousands of residents at St. Mary Parish in Louisiana, USA, received a fake text message alert via a bogus Twitter account warning of a "toxic fume hazard" in the area. Further fanning the flames, a fake YouTube video was also posted showing a masked ISIS fighter standing next to looping footage of an explosion.[89]

False social media reports (e.g. reports of mobile missiles movements, real-time streaming of launches, the deployment of transporter erectors-launchers (TELs), or possible false reports of detonations) might influence the threat sensors of nuclear early warning systems used to inform strategic decision-making. The level of sophistication (i.e. technical know-how and software) needed to execute these kinds of attacks are surprisingly low, with many programs (e.g. voice cloning and GANs software) available at a relatively low-cost (or often free) on the internet. This trend might augur the democratization of ever more sophisticated technology, amplifying the human pathologies (i.e. cognitive heuristics) that underlie this 'information cascade' phenomenon – people's attraction to novel and negative information, and filter bubbles – which explain why technologies like deepfakes are especially adapt to perpetuating destabilizing memes and falsehoods.[90] As explained in chapter 2, during times of high-pressure crises, decision-makers tend to interpret unusual circumstances as threatening, even if an adversary's behavior has not changed. Routine activities (e.g. troop movements) scrutinized in the context of an early warning alert may be considered more menacing than they might otherwise be.[91]

For example, in 2017, South Korean counterintelligence officials received fake mobile and social media alerts with orders for US military and DoD personnel to evacuate the Korean Peninsula.[92] Information attacks such as this suggest that non-state actors, state proxy actors – and perhaps state actors – will inevitably attempt to use social media as a tool of war to provoke nuclear confrontation for political-ideological, religious, or for other malevolent goals; and with increasing levels of sophistication, stratagem, and AI-enhanced subterfuge.[93] AI might also enable states (and non-state actors) to automate, accelerate, and scale synthetic social media accounts and content to support malevolent disinformation operations.[94]

The amplification of false alarms or the creation of false signals by social media (i.e. false positives and false negatives) might also disrupt critical communication channels between commanders and their political leadership, and between allies and adversaries during crisis or conflict.[95] Authoritarian regimes (China, North Korea, Pakistan, or Russia, for example), whose political legitimacy and regime stability is conditioned or legitimized by the general acceptance of official narratives and dogma, tend to become empowered when people's trust in truth (i.e. faith in what they see and hear) is undermined. The vacuum is filled by the opinions of authoritarian regimes and leaders with authoritarian inclinations.[96]

Furthermore, these dynamics may be compounded by human cognitive bias – people's tendency to filter information through the lens of pre-existing beliefs and values.[97] Research has shown that people tend to interpret ambiguous information as consistent with their pre-existing beliefs – dismissing information that contradicts these views – and accept information that allows them to avoid unpleasant choices.[98] Also, humans are considered poorly equipped to intuitively understand probability – which is essential for ranking preferences rationally for crisis decision-making.[99] That is, people tend to misinterpret (or not recognize) randomness and non-linearity and consider the occurrence of unlikely events as virtually impossible.

A motivated authoritarian leader (or non-state actor) would be well-positioned to use AI-augmented tools (e.g. 'fake news' and 'deepfake' propaganda) to exploit this psychological weakness in order to ensure the control and dissemination of false narratives and opinions. Besides, humans exhibit a range of biases that can influence the ways they observe, collect, and process information, which can make them less aware of the reality of a particular situation, and thus more inclined to interpret events through the lens of existing desires, preferences, and beliefs.[100] Strategic decisions made by isolated nuclear-armed authoritarian regimes under these circumstances – especially regimes that believed their survival was threatened – would be a particularly dangerous prospect.

Therefore, without robust, modern, and reliable NC3 structures, decisions to threaten the use of military force and escalate a situation might be premised on false, fabricated, or misperceived narratives.[101] In asymmetric situations, inferior NC3 early warning systems (e.g. ISR systems without long-range sensors, and less able to detect the subtle differences between nuclear and conventional delivery systems) could put leaders under intense pressure to launch a pre-emptive strike due to the perceived use-them-or-lose-them imperatives. Moreover, irregular or opaque communication flow between adversaries may also increase the risk of misperception and miscalculation, and to assume the worst of others' intentions.[102]

Accidental escalation might be set into motion as follows: during heightened tensions or crisis between State A and State B, a third-party actor or terrorist leaks false information (e.g. satellite imagery, 3D models, or geospatial data) into an open-source crowdsourcing platform, about the suspicious movement of State A's nuclear road-mobile TELs.[103] Because of the inability of State B to determine with confidence the veracity of this information, and with mounting public pressures to respond, State B escalates a situation on the (false) belief it is the target of an unprovoked attack.[104] Asymmetries between adversaries' NC3 systems and military capabilities would likely exacerbate the escalation mechanisms illustrated in this fictional scenario.[105] Taken together, increasingly sophisticated GANs, the problem of attribution in cyberspace (see chapter 7), the inherently dual-use nature of AI, the exponentially complex nature of NC3 systems, and the compressed timeframe for strategic decision-making, will continue to lower the threshold for false-flag operations.[106]

Because of the perennial trade-offs between speed, precision, safety, reliability, and trust inherent in cognitive-psychological human–machine interactions, greater emphasis is needed on how, and based on what assumptions, AI systems are designed to replicate human behavior (i.e. preferences, inferences, and judgments).[107] According to former NATO Deputy Secretary General, Rose Gottemoeller, the successful realization of this collaboration will depend on understanding the relative strengths of humans and machines, how they might best function in combination, and to develop "the *right blend of human–machine teams* – the effective integration of humans and machines into our warfighting systems" (emphasis added).[108] In this way, AI can begin to instill trust in how it reaches a particular decision about military force use. Striking this balance will be extremely challenging, however.

Recent research on human–machine teaming in non-military organizations highlights the complexity of optimizing human users' trust in AI systems, with evidence of these users veering from irrational aversion to irrational overconfidence and trust in machines.[109] Thus, as emerging technologies such as AI and autonomy, quantum computing, and big-data analytics are synthesized with and superimposed on states' legacy NC3 systems – at various speeds and degrees of sophistication – new types of errors, distortions, and manipulations (especially involving social media) seem more likely to occur. Critical questions for policy-makers include the following.[110] Should nuclear early warning systems incorporate (or perhaps ignore) social media in their threat assessments? Is there a third party that can provide real-time status of nuclear forces to serve as an independent (and trusted) reference to inform states early warning systems? If so, who should take the lead in creating and verifying it?

To reiterate a central theme of this book: while autonomous nuclear early warning systems may allow planners to identify potential threats faster and more reliably than before, absent human judgment and supervision – coupled with the heightened speed of warfare and the inherent brittleness and vulnerability of ML systems – and the risk of destabilizing accidents and false alarms will likely rise. This analysis speaks to the broader conversation about how divergences between states' nuclear strategy, force structure, and doctrine might affect how they view the use of AI in the nuclear enterprise – or the AI-nuclear dilemma.[111] Today, states face similar contradictions, dilemmas, and trade-offs in the decision about whether or not to integrate AI and autonomy into the nuclear enterprise, as leaders have faced in the quest for strategic stability more generally.[112]

## Conclusion

This chapter considered the risks and trade-offs of increasing the role of machines in the strategic decision-making process, and the impact of synthesizing AI with NC3 early warning systems for the nuclear enterprise. Despite the consensus among nuclear-armed states that decision-making which directly impacts nuclear C2 architecture should not be pre-delegated to machines, once AI demonstrates an ability to engage in strategic planning at a 'superhuman' level, defense planners will likely become more predisposed to view decisions generated by AI algorithms as analogous with (or superior to) those of humans. This predisposition would likely be influenced, or even expedited, by the anthropomorphic tendencies of human psychology. Ultimately, it behooves humans to choose whether or not to abdicate their strategic decision-making authority and to empower AI systems to act on behalf of humankind. It would be a mistake to presume that the absolution of responsibility and accountability necessarily accompanies this transfer of authority.[113]

The chapter found that human agency ensures that AI's role in the nuclear domain is confined to a predominately tactical utility; it might influence strategic decisions that involve nuclear weapons. The distinction between the impact of AI at a tactical and strategic level is *not a binary one*. A particular technology designed to augment autonomous tactical weapons will, nonetheless, ostensibly be making decisions in the use of lethal force that informs and shapes strategic war-faring calculations. This non-binary distinction could risk AI-powered decision support systems substituting the role of critical thinking, empathy, creativity, and intuition of human commanders in the strategic decision-making process. Moreover, the broader issue about control will raise more general ethical questions as to whether tactical AI support systems respond to contingencies in unexplainable and unexpected (i.e., inhuman) ways. Who is responsible and accountable for machines behavior (ethical or otherwise)?

Unlikely as it is that defense planners would explicitly delegate the authority of missile launch platforms, delivery systems, or NC3 to machines, AI technology is expected to be more widely used to support decision-making on strategic nuclear issues – or decision support systems.[114] In short, nuclear states face a trade-off, not only in whether to use AI-enabled decision-support tools, but also how these systems are calibrated to reflect states' risk tolerance (i.e. for false positives vs. false negatives) and confidence in their second-strike capabilities. All things being equal, a state more confident in its ability to retaliate in response to a first strike will be more inclined to design its NC3 systems in ways that do not over-rely on autonomous systems.[115]

This will likely be a double-edged sword for stability in the nuclear enterprise. On the one hand, improvements could increase a state's confidence in their nuclear systems and reassure leaders that an adversary is not planning to launch a pre-emptive strike, thus improving strategic stability. For example, the risk of accidents caused by unauthorized use might be reduced by: bolstering NC3 defenses against physical and cyber-attacks; improving information flow and situational awareness; enhancing nuclear warning and testing systems, thus making accidents caused by mistakes less likely; reducing the risk of human error caused by repetitive tasks and fatigue; and expanding the use of automation in NC3.

On the other hand, these developments could increase escalation risks through two key vectors. First, ML-enhanced cyber-attacks would be considerably more challenging to detect, and thus more effective. Second, in the unlikely event that an attack was successfully detected, threat identification at machine speed would be virtually impossible. In sum, the synthesis of AI – and other advanced emerging technologies such as quantum computing, 5G networks, and big-data analytics – into nuclear early warning systems could further compress the decision-making timeframe and create new network vulnerabilities, thereby eroding crisis stability.

During nuclear brinksmanship, the interaction between machines and human-strategic psychology might increase the risk of misperceiving an adversary's intentions, thus increasing first-strike incentives and undermining crisis stability. Furthermore, AI-supported NC3 systems would be more vulnerable to subversion from cyber-attacks, which could increase the risk of inadvertent escalation – as a result of either human or machine error.[116] To reduce the perceived vulnerability of US NC3 systems to cyber-attacks, the DoD recently proposed a substantial investment to upgrade its NC3 infrastructure.[117]

The nefarious use of AI-enhanced fake news, deepfakes, bots, and other malevolent social media campaigns by non-state actors, terrorists, and state proxies – and perhaps state actors – might also have destabilizing implications on effective deterrence and military planning. In particular, false social media reports (e.g. reports of mobile missiles movements, real-time streaming of launches, the deployment of TELs, or possible false reports of detonations) may influence the threat sensors of nuclear early warning systems used to inform strategic decision-making. *In extremis*, nuclear confrontation could result from false, fabricated, or misperceived narratives.

In combination, an overreliance on AI-enhanced systems (or automation bias) and the risk of false alarms (i.e. especially false positives) in cyberspace might cause states to exaggerate a threat posed by ambiguous or manipulated information, increasing instability. Because nuclear interactions increasingly involve the complex interplay of nuclear and non-nuclear (and non-state) actors, the leveraging of AI in this multipolar context will increasingly place destabilizing pressures on nuclear states.

As emerging technologies like AI are superimposed on states' legacy NC3 systems, more innovative types of errors, distortions, and manipulations seem likely to occur. As described in chapter 2, the framework in which that change could potentially take place is largely consistent with that which has defined most of the nuclear age. Today, states face contradictions, dilemmas, and trade-offs regarding whether or not to integrate AI and autonomy into the nuclear enterprise – just as leaders have faced in the quest for strategic stability, effective deterrence, and enhanced security in a multipolar nuclear world more generally.

What does this analysis mean for the future? How can states (especially great powers) mitigate the potentially escalatory risks posed by AI and steer it to bolster strategic stability as the technology matures? Can we draw on the experiences and best practices of previous emerging technologies to guide us? The final and concluding chapter will consider these questions.

## Notes

1 This chapter is derived in part from an article published in the *Journal of Strategic Studies*, April 30, 2020, copyright Taylor & Francis, available

online: https://doi.org/10.1080/01402390.2020.1759038 (accessed March 10, 2020).

2 In 2011, Commander-in-Chief of the Russian Strategic Rocket Forces, General S. Karakayev, confirmed in an interview with one of the central Russian newspapers that *Perimeter* – also known as Russia's Dead Hand Doomsday Weapon – exists and continues to be on combat duty. The system's characteristics and capabilities are unknown, however. See Eric Schlosser, *Command and Control* (New York: Penguin Group, 2014); Richard Rhodes, *Arsenals of Folly: The Making of the Nuclear Arms Race* (London: Simon & Schuster, 2008); and Ryabikhin Leonid, "Russia's NC3 and Early Warning Systems," *Tech4GS*, July 11, 2019, www.tech4gs.org/nc3-systems-and-strategic-stability-a-global-overview.html (accessed December 10, 2019).

3 During the 1950s, Soviet bombers would take hours to reach the US. The missile age compressed this timeframe to around 30 minutes for a land-based intercontinental ballistic missile, and about 15 minutes for a submarine-launched ballistic missile. "US DoD Strategic Command, US Strategic Command & US Northern Command SASC Testimony," March 1, 2019, www.stratcom.mil/Media/Speeches/Article/1771903/us-strategic-command-and-us-northern-command-sasc-testimony/ (accessed December 10, 2019).

4 Geoffrey Forden, Pavel Podvig, and Theodore A. Postol, "False Alarm, Nuclear Danger," *IEEE Spectrum*, 37, 3 (2000), pp. 31–39.

5 Modern NC3 systems include: early warning satellites, radars, and sensors; hydroacoustic stations, and seismometers facilities to collect and interpret early warning information; fixed and mobile networked command posts; and a communications infrastructure that includes landlines, satellite links, radars, radios, and receiving terminals in ground stations and aboard strike vehicles. See Amy Woolf, *Defense Primer: Command and Control of Nuclear Forces* (Washington, DC: Congressional Research Service), December 11, 2018, p. 1.

6 For example, DARPA's Professional, Educated, Trained, and Empowered AI-enabled virtual assistant gathers and collates information, as well as liaising and executing orders from commanders. Peter W. Singer, *Wired for War: The Robotics Revolution and Conflict in the 21st Century* (New York: Penguin Group, 2009), p. 359.

7 Jon R. Lindsay, "Cyber Operations and Nuclear Weapons," *Tech4GS Special Reports*, June 20, 2019, https://nautilus.org/napsnet/napsnet-special-reports/cyber-operations-and-nuclear-weapons/ (accessed December 10, 2019).

8 Jeffrey Larsen, "Nuclear Command, Control, and Communications: US Country Profile," *Tech4GS Special Reports*, August 22, 2019, https://nautilus.org/napsnet/napsnet-special-reports/nuclear-command-control-and-communications-us-country-profile/ (accessed December 10, 2019).

9 See Bruce Blair, *Strategic Command and Control: Redefining the Nuclear Threat* (Washington DC: Brookings Institution, 1985); Shaun Gregory, *The Hidden Cost of Deterrence: Nuclear Weapons Accidents* (London: Brassey's, 1990); Scott D. Sagan, *The Limits of Safety: Organizations, Accidents, and Nuclear Weapons* (Princeton, NJ: Princeton University Press, 1995); and Eric

Schlosser, *Command and Control: Nuclear Weapons, the Damascus Accident, and the Illusion of Safety* (New York: Penguin, 2014).

10 The 'cognitive revolution' (related to the field of cognitive science) encompasses rich literature, including the philosophy of the mind, neuroscience, and AI, to name but a few. See Douglas R. Hofstadter and Daniel C. Dennett, *The Mind's I* (New York: Basic Books, 2001).

11 Wendell Wallach and Colin Allen, *Moral Machines* (New York: Oxford University Press, 2009), p. 40.

12 DeepMind's AlphaStar victory represented a technical milestone in several ways: (1) using game-theory logic to find ways to improve and expand its boundaries continually; (2) unlike games such as chess or Go, operating in an imperfect information situation; (3) performing long-term planning in real-time; and (4) controlling large and complex possibilities with combinatorial space possibilities (i.e. hundreds of units, personnel, and buildings), in real-time. AlphaStar Team, "Alphastar: Mastering the Real-Time Strategy Game StarCraft II," *DeepMind Blog*, January 24, 2019, https://deepmind.com/blog/article/alphastar-mastering-real-time-strategy-game-starcraft-ii (accessed March 10, 2020).

13 Cade Metz, "In Two Moves, AlphaGo and Lee Sedol Redefined the Future," *Wired*, March 16, 2016, www.wired.com/2016/03/two-moves-alphago-lee-sedol-redefined-future/ (accessed December 10, 2019).

14 AI's technical milestones in a virtual environment would, however, unlikely be replicated in stochastic (i.e. randomly determined) and complex systems like NC3. See AlphaStar Team, "Alphastar: Mastering the Real-Time Strategy Game StarCraft II," *DeepMind Blog*.

15 'Theory of the mind' allows humans to understand that other humans may hold intentions and beliefs about a situation or action that differs from what they believe. It enables humans to make predictions about the behavior and intentions of others. Brittany N. Thompson, "Theory of Mind: Understanding Others in a Social World," *Psychology Today*, July 3, 2017, www.psychologytoday.com/us/blog/socioemotional-success/201707/theory-mind-understanding-others-in-social-world (accessed December 10, 2019).

16 See Kareem Ayoub and Kenneth Payne, "Strategy in the age of Artificial Intelligence," *Journal of Strategic Studies* 39, 5–6 (2016), pp. 793–819, p. 814; and James Johnson, "Delegating Strategic Decision-Making to Machines: Dr. Strangelove Redux?" *Journal of Strategic Studies* (2020), www.tandfonline.com/doi/abs/10.1080/01402390.2020.1759038 (accessed February 5, 2021).

17 Carl von Clausewitz, *On War*, trans. Michael Howard and Peter Paret (Princeton, NJ: Princeton University Press, 1976), p. 140.

18 Vincent Boulanin (ed.), *The Impact of Artificial Intelligence on Strategic Stability and Nuclear Risk Vol. I Euro-Atlantic Perspectives* (Stockholm: SIPRI Publications, May 2019), p. 56.

19 For example, see Linda J. Skitka, Kathleen L. Mosier, and Mark Burdick, "Does Automation Bias Decision-Making?" *International Journal of Human-Computer Studies*, 51, 5 (1999), pp. 991–1006; and Mary L. Cummings,

"Automation Bias in Intelligent Time-Critical Decision Support Systems," *AIAA 1st Intelligent Systems Technical Conference* (2004), pp. 557–562.

20 Edward Geist and Andrew Lohn, *How Might Artificial Intelligence Affect the Risk of Nuclear War?* (Santa Monica, CA: RAND Corporation, 2018), p. 17.

21 David Watson, "The Rhetoric and Reality of Anthropomorphism in Artificial Intelligence," *Minds and Machines* 29 (2019), pp. 417–440, p. 434.

22 See Marilynn B. Brewer and William D. Crano, *Social Psychology* (New York: West Publishing Co. 1994).

23 Clifford Nass, B.J. Fogg, and Youngme Moon, "Can Computers be Teammates?" *International Journal of Human-Computer Studies* 45 (1996), pp. 669–678.

24 Sharing system monitoring tasks and decision-making functions with machines might also have psychological parallels between humans sharing similar tasks. A considerable body of psychology research suggests that people tend to expend less effort when working collectively than working alone. Steven J. Karau and Kipling D. Williams, "Social Loafing: A Meta-Analytic Review and Theoretical Integration," *Journal of Personality and Social Psychology*, 65 (1993) pp. 681–706.

25 One area of encouragement from recent studies on automation bias and errors is that participants who were made explicitly aware and received training on automation bias were less likely to make certain classes of errors. Linda J. Skitka, Kathleen L. Mosier, and Mark Burdick, "Does Automation Bias Decision-Making?" *International Journal of Human-Computer Studies* 51, 5 (1999), pp. 991–1006.

26 Parasuraman, Raja, and Victor Riley, "Complacency and bias in human use of automation: An attentional integration," *Human Factors*, 52, 3 (2010), pp. 381–410.

27 Ibid.

28 See Raja Parasuraman and Victor Riley, "Humans and Automation: Use, Misuse, Disuse, Abuse," *Human Factors*, 39, 2 (June 1997), pp. 230–253.

29 For example, during the Cold War, the Soviets developed a computer program known as VRYAN (a Russian acronym for 'Surprise Nuclear Missile Attack') to notify the leadership in Moscow of a pre-emptive US nuclear strike. However, the data used to feed the system was often biased and, thus, propelled a feedback loop, which heightened the Kremlin's fear that the US was pursuing first-strike superiority. President's Foreign Intelligence Advisory Board, "The Soviet 'War Scare,'" February 15, 1990, vi, 24 et seq, https://nsarchive2.gwu.edu/nukevault/ebb533-The-Able-Archer-War-Scare-Declassified-PFIAB-Report-Released/ (accessed March 10, 2020).

30 John J. Mearsheimer, "The Gathering Storm: China's Challenge to US Power in Asia," *The Chinese Journal of International Politics*, 3, 4 (Winter 2010), pp. 381–396.

31 Keren Yarhi-Milo, *Knowing the Adversary* (Princeton, NJ: Princeton University Press, 2014), p. 250.

32 For example, see Yuan Yi, "The Development of Military Intelligentization Calls for Related International Rules," *PLA Daily*, October 16, 2019, http://military.workercn.cn/32824/201910/16/191016085645085.shtml (accessed March 10, 2020).
33 Brian W. Everstine, "DOD AI Leader Wants Closer Collaboration With NATO," *Airforce Magazine*, January 15, 2020, www.airforcemag.com/dod-ai-leader-wants-closer-collaboration-with-nato/ (accessed January 17, 2020).
34 "Generating Actionable Understanding of Real-World Phenomena with AI," *DARPA*, January 4, 2019, www.darpa.mil/news-events/2019-01-04 (accessed December 10, 2019).
35 The US, for instance, does not field communication satellites exclusively for nuclear operations. Curtis Peebles, *High Frontier: The US Air Force and the Military Space Program* (Washington, DC: Air Force Historical Studies Office, January 1997), pp. 44–52.
36 NC3 systems must meet the 'always never' criteria, which is critical to deterrence. Nuclear weapons must always work when tasked to and never go off accidentally or without proper authorization. Moreover, NC3 systems must, under all circumstances, be able to execute a lawful order to employ nuclear force (known as 'positive control'). At the same time, NC3 must never allow the nuclear force to be used accidentally or by an illegitimate authority (known as 'negative control'). Larsen, "Nuclear Command, Control, and Communications: US Country Profile," pp. 10–11.
37 The historical record demonstrates that human strategists rarely have a clear idea from the outset of what they are seeking to achieve through pursuing a particular strategic path, less still how these goals might be realized. Kenneth Payne, "Fighting on Emotion and Conflict Termination," *Cambridge Review of International Affairs* 28/3 (August 2015), pp. 480–497.
38 Peter Hayes, Binoy Kampmark, Philip Reiner, and Deborah Gordon, "Synthesis Report, NC3 Systems, and Strategic Stability: A Global Overview," *Tech4GS Special Reports*, May 6, 2019, www.tech4gs.org/nc3-systems-and-strategic-stability-a-global-overview.html (accessed December 10, 2019).
39 Kenneth Payne, *Strategy, Evolution, and War: From Apes to Artificial Intelligence* (Washington, DC: Georgetown University Press), p. 183.
40 Ibid.
41 US Department of Defense, *Nuclear Posture Review* (Washington DC: US Department of Defense, February 2018), pp. 57–58.
42 Today, for example, big data already enables probabilistic prediction of people's political attitudes, activism, and their violent tendencies. See Jakob Bæk Kristensen et al., "Parsimonious Data: How a Single Facebook Like Predicts Voting Behavior in Multiparty Systems," *PLOS One* 12, 9 (2017); Petter Bae Brandtzaeg, "Facebook is No 'Great Equalizer': A Big Data Approach to Gender Differences in Civic Engagement Across Countries," *Social Science Computer Review* 35, 1 (2017), pp. 103–125; and Andrea Pailing, Julian Boon, and Vincent Egan, "Personality, the Dark Triad and Violence," *Personality and Individual Differences* 67 (September 2014), pp. 81–86.

43 Jia Daojin and Zhou Hongmei, "The Future 20–30 Years Will Initiate Military Transformation," *China Military Online*, June 2, 2016, www.81.cn/jmywyl/2016-06/02/content_7083964.htm (accessed December 10, 2019); and Yang Feilong and Li Shijiang "Cognitive Warfare: Dominating the Era of Intelligence," *PLA Daily*, March 19, 2020, www.81.cn/theory/2020-03/19/content_9772502.htm (accessed December 10, 2019).

44 "Evidence Reasoning and Artificial Intelligence Summit Forum," December 26, 2017, http://xxxy.hainnu.edu.cn/html/2018/xyxw_0111/1003.html (accessed December 10, 2019).

45 For example, see "Chinese Commercial Space Start-Ups Launch Two AI Satellites in a Hundred Days," *Global Times*, November 26, 2018, http://smart.huanqiu.com/ai/2018-11/13645096.html?agt=15422 (accessed March 10, 2020).

46 From the open sources, no unambiguous evidence has emerged to suggest China has – or plans to use in the near-future – AI to augment its NC3 systems.

47 Wendell Wallach and Colin Allen, *Moral Machines* (New York: Oxford University Press, 2009), chapters 3 and 4.

48 Though computer programs, simulations, and data analysis are already used to inform human defense planners, AI operating at superhuman speed and performing increasingly complex tasks is likely to accelerate this trend, as Google's AlphaGo now-infamous defeat of the Go world champion attested. Darrell Etherington, "Google's AlphaGo AI Beats the World's Best Human Go Player," *TechCrunch*, May 23, 2017. As of August 15, 2017: https://techcrunch.com/2017/05/23/googles-alphago-ai-beats-the-worldsbest-human-go-player/ (accessed December 10, 2019).

49 Yuna Huh Wong et al., *Deterrence in the Age of Thinking Machines* (Santa Monica, CA: RAND Corporation, 2020), www.rand.org/pubs/research_reports/RR2797.html (accessed March 10, 2020).

50 Signaling during wartime has also been a challenging balancing act, especially between different strategic cultures. For example, during the Vietnam War, strategic game theory influenced US bombing decision planning, but this approach underestimated the role of uncertainty and unpredictable human psychology during warfare. Singer, *Wired for War*, pp. 305–306.

51 The term 'rationality-of-irrationality' refers to a class of bargaining, negotiating tactics, or escalation scenarios, where there exists a rational advantage to be gained from irrational behavior – or the expectation of irrational behavior – to enhance deterrence. Herman Kahn, *On Escalation: Metaphors and Scenarios* (New York: Praeger, 1965), pp. 57–58.

52 Thomas C. Schelling, *Arms and Influence* (New Haven, CT: Yale University Press, 1966), pp. 115–116.

53 Psychologists have shown that people are generally averse to ambiguity; thus, they are more willing to accept smaller gains to avoid being placed in an ambiguous situation. See Stefan T. Trautmann and Gijs van de Kuilen, "Ambiguity attitudes," in *The Wiley Blackwell Handbook of Judgment and Decision Making*, ed. Gideon Keren and George Wu (Chichester: John Wiley & Sons, 2015).

54 Mary L. Cummings, "Automation Bias in Intelligent Time-Critical Decision Support Systems," *AIAA 1st Intelligent Systems Technical Conference*, 2004, pp. 557–562 https://arc.aiaa.org/doi/abs/10.2514/6.2004-6313 (accessed March 10, 2020).
55 For the foreseeable future, AI ML – and in particular the deep-learning subset – used to predict adversaries' intentions will be highly correlative and dependent on a range of engineering factors that rely on human-compiled data on historical patterns, and the parameters of select modeling practices. Benjamin M. Jensen, Christopher Whyte, and Scott Cuomo, "Algorithms at War: The Promise, Peril, and Limits of Artificial Intelligence," *International Studies Review* (June 2019), pp. 526–550, p. 15.
56 David Whetham and Kenneth Payne, "AI: In Defence of Uncertainty," *Defence in Depth*, December 9, 2019, https://defenceindepth.co/2019/12/09/ai-in-defence-of-uncertainty/ (accessed March 10, 2020).
57 Lora Saalman, "Lora Saalman on How Artificial Intelligence Will Impact China's Nuclear Strategy," *The Diplomat*, November 7, 2018, https://thediplomat.com/2018/11/lora-saalman-on-how-artificial-intelligence-will-impact-chinas-nuclear-strategy/ (accessed March 10, 2020).
58 On the risks of states exaggerating an adversary's capabilities and misinterpreting their strategic intentions see Robert L. Jervis, *The Logic of Images in International Relations* (New York: Columbia University Press, 1989); Barton Whaley, "Covert Rearmament in Germany, 1919–1939: Deception and Misperception," *Journal of Strategic Studies*, 5, 1 (March 1982), pp. 3–39; and Keren Yarhi-Milo, "In the Eye of the Beholder: How Leaders and Intelligence Communities Assess the Intentions of Adversaries," *International Security*, 38, 1 (Summer 2013), pp. 7–51.
59 There is a range of different ways to subvert AI systems, and given the embryonic nature of AI cyber-defense, the offense is likely to have the upper hand in this domain for the near future. Hyrum S. Anderson et al., "Evading Machine Learning Malware Detection," *blackhat.com*, July 20, 2017, www.blackhat.com/docs/us-17/thursday/us-17-Anderson-Bot-Vs-Bot-Evading-Machine-Learning-Malware-Detection-wp.pdf (accessed December 10, 2019); and Battista Biggio, Blaine Nelson, and Pavel Laskov, "Poisoning Attacks Against Support Vector Machines," *Proceedings of the 29th International Conference on Machine Learning*, July 2012, pp. 1467–1474 https://arxiv.org/pdf/1206.6389.pdf (accessed December 10, 2019).
60 James Johnson, "The AI-Cyber Nexus: Implications for Military Escalation, Deterrence and Strategic Stability," *Journal of Cyber Policy*, 4, 3 (2019), pp. 442–460.
61 Nicolas Papernot, Patrick McDaniel, and Ian Goodfellow, "Transferability in Machine Learning: From Phenomena to Black-Box Attacks Using Adversarial Samples," *arXiv*, May 24, 2016, https://arxiv.org/abs/1605.07277 (accessed March 10, 2020).
62 During the formulation of Russia's most recent military doctrine, Russian strategists proposed that an attack on Russia's early warning systems

would be interpreted as a sign of an impending nuclear attack. This proposal was not, however, included in the final version. Ryabikhin Leonid, "Russia's NC3 and Early Warning Systems," *Tech4GS* www.tech4gs.org/uploads/1/1/1/5/111521085/russia_nc3_tech4gs_july_11–2019_3.pdf p. 10 (accessed December 10, 2019).

63 The historical record is replete with examples of false alerts and warnings from satellite and over-the-horizon radars. See Andrew Futter, *Hacking the Bomb: Cyber Threats and Nuclear Weapons* (Washington DC: Georgetown University Press, 2018), chapter 2.

64 Johnson, "Delegating Strategic Decision-Making to Machines: Dr. Strangelove Redux?"

65 Clausewitz, *On War*, pp. 112–117.

66 'Mission command' is a modern military concept, especially popular with British and US forces, which relies on the creativity, intuition, and expertise of subordinates closer to action to implement a commander's intent. See Col. (Ret.) James D. Sharpe Jr. and Lt. Col. (Ret.) Thomas E. Creviston, "Understanding mission command," *US Army*, July 10, 2013, www.army.mil/article/106872/understanding_mission_command (accessed December 10, 2019).

67 A notable example of the failure of 'mission command' was the ill-timed ICBM test at the US Vandenberg Air Force Base during the height of the 1962 Cuban Missile Crisis. Stephen J. Cimbala, *The Dead Volcano: The Background and Effects of Nuclear War Complacency* (New York: Praeger, 2002), p. 66. For more on these debates, see Stephen D. Biddle, *Military Power: Victory and Defeat in Modern Warfare* (Princeton, Princeton University Press, 2004). Peter W. Singer, "Robots and the Rise of Tactical Generals," *Brookings*, March 9, 2009, www.brookings.edu/articles/robots-and-the-rise-of-tactical-generals/ (accessed December 10, 2019).

68 Studies have demonstrated over 180 types of human cognitive biases and limitations (e.g. working memory, attention, confirmation bias, and loss aversion). Buster Benson, "Cognitive Bias Cheat Sheet," *Better Humans*, September 1, 2016, https://medium.com/better-humans/cognitive-bias-cheat-sheet-55a4724 76b18 (accessed December 10, 2019).

69 In early tests of RAID between 2004 and 2008, the system performed with greater accuracy and speed than human planners. Alexander Kott and Michael Ownby, "Tools for Real-Time Anticipation of Enemy Actions in Tactical Ground Operations," *Defense Technical Information Center*, June 2005 https://apps.dtic.mil/dtic/tr/fulltext/u2/a460912.pdf (accessed December 10, 2019).

70 Sandra Erwin, "BAE wins DARPA Contract to Develop Machine Learning Technology for Space Operations," *Spacenews*, August 13, 2019, https://spacenews.com/bae-wins-darpa-contract-to-develop-machine-learning-technology-for-space-operations/ (accessed December 10, 2019).

71 In contrast to relatively structured institutional decision-making settings, such as voting and consumer behavior, a conflict situation typically encompasses

a substantially greater set of actors interacting in surprising and, by definition, rule-breaking ways. See "Predicting Armed Conflict: Time to Adjust Our Expectations?" Lars-Erik Cederman, *Science*, 355 (2017), pp. 474–476.

72 AI-assisted predictive policing has already had some notable successes in combating crime. For example, in Los Angeles, police claim to have reduced burglaries by 33 percent and violent crime by 21 percent through AI-driven predictive policing, and Chicago has established an algorithmically derived list of individuals considered most likely to commit crimes. Also, in the UK, a predictive policing platform called PredPol has been designed to predict where future crime will likely occur based on data-sets of anonymized victim reports. Similarly, Japan, Singapore, and, most notably, China, have implemented similar systems. Keith Dear, "Artificial Intelligence and Decision-Making," *The RUSI Journal*, 164, 5–6 (2019), pp. 18–25; and Yang Feilong and Li Shijiang "Cognitive Warfare: Dominating the Era of Intelligence," *PLA Daily*, March 19, 2020, www.81.cn/theory/2020-03/19/content_9772502.htm (accessed April 10, 2020).

73 For example, in 1983, a malfunctioning Soviet early warning system led to the "detection" of a nonexistent US attack.

74 Experts believe that once deepfakes become widespread, even having a human involved may not suffice to determine the veracity or source of a specific post or site. Nautilus Institute, Technology for Global Security, Preventive Defense Project, "Social Media Storms and Nuclear Early Warning Systems: A Deep Dive and Speed Scenario Workshop," *NAPSNet Special Reports*, January 8, 2019, https://nautilus.org/napsnet/napsnet-special-reports/social-media-storms-and-nuclear-early-warning-systems-a-deep-dive-and-speed-scenarios-workshop-report/ (accessed December 10, 2019).

75 Paige Gasser, Rafael Loss, and Andrew Reddie, "Assessing the Strategic Effects of Artificial Intelligence – Workshop Summary," Center for Global Security Research (Livermore, CA: Lawrence Livermore National Laboratory), September 2018, p. 9. https://cgsr.llnl.gov/content/assets/docs/Final_AI_Workshop_Summary.pdf (accessed December 10, 2019).

76 A recent workshop hosted by the UK-based think-tank IISS demonstrated that malign manipulation of input data received by early warning systems might not only subvert the output of AI systems in specific situations but also undermine the reliability of an entire algorithm network environment. In particular, if an attack was executed during the 'training' phase for such programs (e.g. pattern recognition or intelligence gathering and analysis software). Mark Fitzpatrick, "Artificial Intelligence and Nuclear Command and Control," *Survival*, 61, 3 (2019), pp. 81–92.

77 For example, in 2019, non-state actors used AI voice-mimicking software to generate a fake recording of a British energy executive to conduct the world's first reported AI-enabled theft. Drew Harwell, "An Artificial-Intelligence First: Voice-Mimicking Software Reportedly Used in a Major Theft," *Washington Post*, September 4, 2019, www.washingtonpost.com/technology/2019/09/04/an-artificial-intelligence-first-voice-mimicking-software-reportedly-used-major-theft/ (accessed December 10, 2019).

78 Disinformation in deception and misinformation campaigns is a familiar aspect of warfare, perhaps most famously demonstrated by the efforts of the Allies during World War II's *Operation Bodyguard* to mislead the Axis regarding the location of what became the D-Day invasion of 1944. Jamie Rubin, "Deception: The Other 'D' in D-Day," *NBC News*, June 5, 2004, www.nbcnews.com/id/5139053/ns/msnbc-the_abrams_report/t/deception-other-d-dday/#.WvQt5NMvyT8 (accessed December 10, 2019).

79 Russia, Pakistan, and perhaps China have reportedly indicated a possible willingness to use limited nuclear strikes to end a conventional war that is losing. Michael C. Horowitz, Paul Scharre, and Alexander Velez-Green, "A Stable Nuclear Future? The Impact of Autonomous Systems and Artificial Intelligence," December 2019, *arXix*, https://arxiv.org/pdf/1912.05291.pdf, pp. 32–33 (accessed March 10, 2020).

80 For example, even if the malware detected in an attack was only capable of espionage, a target might fear that it also contained a "kill switch," disabling its early warning system after activation.

81 Saalman, "Fear of False Negatives: AI and China's Nuclear Posture."

82 The uncertainty and ambiguity of states' co-mingled nuclear and non-nuclear weapon systems can lead to mischaracterization that causes false positives or false negatives. Misinterpretation of these ambiguous capabilities by intelligence communities in situations of imperfect (or manipulated) information can exacerbate these dynamics. For example, during the Cuban Missile Crisis, the US was confronted with multiple ambiguous Soviet weapon systems and assumed that two types of aircraft were nuclear-armed, even though there was minimal evidence to support their assessment. See James M. Acton, *Is this a Nuke? Pre-Launch Ambiguity and Inadvertent Escalation* (New York: Carnegie Endowment for International Peace, 2020), p. 3.

83 "Is Launch Under Attack Feasible?" *Nuclear Threat Initiative*, August 4, 2016, www.nti.org/newsroom/news/new-interactive-launch-under-attack-feasible/ (accessed December 10, 2019).

84 Charles Perrow, *Normal Accidents: Living with High-Risk Technologies* (Princeton, NJ: Princeton University Press, 1999), p. 259.

85 Thomas C. Schelling, *Arms and Influence* (New Haven, CT: Yale University Press, 1966), p. 234.

86 For example, the China Aerospace Science and Industry Corporation has become active in the development of dual-use networks for quantum communications, which may be used to transmit classified military information between command and control centers and military units during combat. Raymond Wang, "Quantum Communications and Chinese SSBN Strategy," *The Diplomat*, November 4, 2017, https://thediplomat.com/2017/11/quantumcommunications-and-chinese-ssbn-strategy/ (accessed March 10, 2020).

87 Gasser, Loss, and Reddie, "Assessing the Strategic Effects of Artificial Intelligence – Workshop Summary," p. 10.

88 Between August 2017 and January 2018, six instances of social media playing a role in nuclear-prone conflicts occurred in the Asia-Pacific region alone. See

Nautilus Institute, Technology for Global Security, Preventive Defense Project, "Social Media Storms and Nuclear Early Warning Systems," p. 1.

89 Adrian Chen, "The Agency," *New York Times Magazine*, June 2, 2015, www.nytimes.com/2015/06/07/magazine/the-agency.html (accessed December 10, 2019).

90 Danielle K. Citron and Robert Chesney, "Deep Fakes: A Looming Challenge for Privacy, Democracy, and National Security," 107 *California Law Review* 1753 (2019), https://papers.ssrn.com/sol3/papers.cfm?abstract_id=3213954 (accessed December 10, 2019).

91 Robert Jervis, *How Statesmen Think: The Psychology of International Politics* (Princeton, NJ: Princeton University Press, 2017), p. 222.

92 Dan Lamothe, "US Families Got Fake Orders to Leave South Korea. Now Counterintelligence is Involved," *The Washington Post*, September 22, 2017, www.washingtonpost.com/news/checkpoint/wp/2017/09/22/u-s-families-got-fake-orders-to-leave-south-korea-now-counterintelligence-is-involved/ (accessed December 10, 2019).

93 AI systems can track individuals' or groups' online habits, knowledge, and preferences to calibrate specific messages (i.e. propaganda) to maximize the impact on that individual (or group) and minimize the risk of this information being questioned. The information can then be used by AI systems in real-time to determine the messages' influence, thus learning to become more effective in its task. Stuart Russell, *Human Compatible: Artificial Intelligence and the Problem of Control* (New York: Viking Press, 2019), p. 105.

94 Alina Polyakova, "Weapons of the Weak: Russia and AI-Driven Asymmetric Warfare," *Brookings*, November 15, 2018, www.brookings.edu/research/weapons-of-the-weak-russia-and-ai-driven-asymmetric-warfare/ (accessed December 10, 2019).

95 The use of false information, false signals, and elaborate spoofing between adversaries is not an entirely new phenomenon. For example, during World War II, the Japanese sent false radio traffic signals to the US to create the impression that certain ships were maneuvering near Japan's mainland. Japanese commanders also sent false war plans for Chinese targets, which were then altered at the last minute to align them with the South-Eastern movement. Roberta Wohlstetter, *Pearl Harbor: Warning and Decision* (Palo Alto, CA: Stanford University Press, 1962), pp. 393–394.

96 In contested information environments, errors in early warning systems and decision-making processes are characterized by a lack of data, ambiguous indicators, mixed signals, and conflicting sensor data inputs. Amidst this complexity and ambiguity, false signals frequently occur (and are even expected), not least because sensor systems may not be cross calibrated to provide cross-checking confirmation. Nautilus Institute, Technology for Global Security, Preventive Defense Project, "Social Media Storms and Nuclear Early Warning Systems," p. 12.

97 Jervis, *Perception and Misperception in International Politics*, pp. 117–202.

98 Michela Del Vicario et al., "Modeling Confirmation Bias and Polarization," *Sci*

Rep (2017), www.nature.com/articles/srep40391 (accessed February 5, 2021).
99 Nassim N. Taleb, *Fooled by Randomness: The Hidden Role of Chance in Life and the Markets*, 2nd ed. (London: Penguin, 2004).
100 Cordelia Fine, *A Mind of Its Own: How Your Brain Distorts and Deceives* (New York: W. W. Norton & Company, 2008).
101 Several nuclear weapons states – North Korea, Pakistan, and India – have much less capable early warning systems than the US, using fewer satellites and other long-distance sensors such as radars with limited coverage. North Korea, for example, does not possess any long-range sensor systems. Nautilus Institute, Technology for Global Security, Preventive Defense Project, "Social Media Storms and Nuclear Early Warning Systems," p. 6.
102 Robert Jervis and Mira Rapp-Hooper, "Perception and Misperception on the Korean Peninsula," *Foreign Affairs*, April 5, 2018, www.foreignaffairs.com/articles/north-korea/2018-04-05/perceptionand-misperception-korean-peninsula (accessed December 10, 2019).
103 From an intelligence standpoint, nuclear solid-fuel missiles and tracked TELs reduce ISR systems' ability to detect launch preparation signs. Moreover, solid fuel also increases the speed at which missiles can be launched and reduces the number of support vehicles to support an operation.
104 As demonstrated in chapter 7, alternative outcomes from this fictional scenario are possible. For example, counter-AI systems might uncover the source or false nature of the leak before it can do severe damage. State A might also be able to assure State B through backchannel or formal diplomatic communications of this falsehood. While social media platforms have had some success in slowing down users' ability to orchestrate manipulative and dangerous campaigns, once these operations (e.g. deepfakes and bots) go viral, the ability to curtail them becomes inexorably problematic – for human operators or machines.
105 Other information related to weapons of mass destruction that would likely be considered equally escalatory might include intelligence about the movement of support vehicles delivering liquid-fuel to prepare a liquid-fueled missile for launch, spikes in radiation levels, or the detection of chemical warfare agents.
106 See Herbert Lin, "Escalation Dynamics and Conflict Termination in Cyberspace," *Strategic Studies Quarterly*, 6, 3 (Fall, 2012), pp. 46–70.
107 As a counterpoint, psychologists have argued that because interactions between humans and machines are determined primarily by a machine's ability to fulfill a specific task – and not attributes in human-to-human interactions such as benevolence and integrity – trust is, therefore, more straightforward. See Kevin A. Hoff and Masooda N. Bashir, "Trust in Automation: Integrating Empirical Evidence on Factors that Influence Trust," *Human Factors*, 57, 3 (2015), pp. 407–434.
108 Remarks by NATO Deputy Secretary General Rose Gottemoeller at the Xiangshan Forum in Beijing, China, October 25, 2018, NATO Newsroom, Speeches & Transcripts Section, www.nato.int/cps/en/natohq/opinions_160121.htm (accessed December 10, 2019).

109 Roger C. Mayer et al., "An Integrative Model of Organizational Trust," *Academy of Management Review*, 20, 3 (1995), pp. 709–734; and Berkeley Dietvorst et al., "Algorithm Aversion: People Erroneously Avoid Algorithms After Seeing Them Err," *Journal of Experimental Psychology*, 144, 1 (2014), pp. 114–126.
110 Nautilus Institute, Technology for Global Security, Preventive Defense Project, "Social Media Storms and Nuclear Early Warning Systems," p. 30.
111 See Stephen D. Biddle and Robert Zirkle, "Technology, Civil–Military Relations, and Warfare in the Developing World," *Journal of Strategic Studies*, 19, 2 (1996), pp. 171–212; and Johnson, "Delegating Strategic Decision-Making to Machines: Dr. Strangelove Redux?"
112 Political psychology literature has demonstrated that humans tend to avoid trade-offs that involve complex or ambiguous moral and ethical choices, both for cognitive and functional reasons. See Jervis, *Perception and Misperception in International Politics*, pp. 128–142; and Alan Fiske and Philip Tetlock, "Taboo Trade-Offs: Reactions to Transactions that Transgress the Spheres of Justice," *Political Psychology* 18 (June 1997), pp. 255–297.
113 Watson, "The Rhetoric and Reality of Anthropomorphism in Artificial Intelligence," pp. 434–435.
114 The Soviet experience during the 1980s suggests that the possibility of movement in that direction should not be discounted. See Pavel Podvig, "History and the Current Status of the Russian Early-Warning System," *Science and Global Security*, 10, 1 (2002), pp. 21–60.
115 Russia's willingness to contemplate the use of a long-duration, underwater, uninhabited nuclear delivery vehicle – called the Poseidon or Status-6 – demonstrates that states' fear of military (conventional or nuclear) inferiority can manifest in incentives to pursue greater autonomy. Horowitz et al., "A Stable Nuclear Future?" p. 18.
116 For example, the US destruction of the Chinese embassy in Belgrade in 1999 illustrates that accidents in the military arena have broader, long-term geopolitical and geostrategic implications.
117 US Department of Defense, *Nuclear Posture Review*.

# Conclusion: managing an AI future

This book has advanced the case for narrow AI as a fundamentally destabilizing force, which could increase the risk of nuclear war. It has explained how, left unchecked, the uncertainties created by the rapid proliferation and diffusion of AI into advanced weapon systems will become a significant source of future instability and great-power (especially US–China) strategic competition. The book has conceptualized recent technological developments in AI with the broader spectrum of emerging technologies – robotics and autonomy, cyberspace, hypersonic weapons, 5G networks, and quantum communications – and analyzed the impact of these trends for future warfare between nuclear states. Anticipating and preparing for the consequences of AI has already become a critical – yet underappreciated – task for international security, defense planning, and statecraft.

The book's four overarching findings help square the circle between the book's theoretical and empirical parts. These themes were weaved through the substantive chapters in Part II, which considered the implications of AI-augmented capabilities for strategic stability. First, AI does *not exist in a vacuum*. That is, as a stand-alone portfolio of applications, military AI has a limited direct impact on strategic stability.[1] Instead, AI is best viewed as a potentially powerful enabler and force multiplier of a broad portfolio of capabilities, rather than a 'weapon' per se, which might mutually reinforce the destabilizing effects of existing advanced capabilities. Further, the interplay of AI with nuclear weapons and a broad spectrum of strategic non-nuclear weapons – especially conventional counterforce capabilities – will likely exacerbate the erosion of the survivability of nuclear arsenals associated with the digital (or computer) revolution described in chapter 2.[2]

Chapter 5 explained how AI-enhanced intelligence, surveillance, and reconnaissance (ISR), automatic target recognition, and terminal guidance capabilities might begin to erode the critical deterrence role of mobile intercontinental ballistic missiles and ballistic missile submarines. From a technical perspective, as explained in chapter 1, AI has not yet evolved to a point where it can credibly threaten the survivability of a state's second-strike capability. For now, the development of AI (especially the machine learning (ML) subset) will likely be more prosaic and evolutionary than revolutionary. In sum, AI-enabled, and enhanced capabilities will have a more significant impact (positive or negative) on strategic stability, than the sum of its parts. The long-term strategic implications of adopting military AI may be greater than those of any specific military operation or capability.

Second, AI's impact on stability, deterrence, and escalation will be determined as much (or more so) by states' perception of its functionality than what it is – technically, tactically, or operationally – capable of doing. Thus, the effects of AI in a military context will continue to have a strong cognitive element (or human agency), thereby increasing the risk of inadvertent or accidental escalation caused by misperception or miscalculation. Put another way, the perception of a particular AI-augmented capability's unpredictability will likely become a more significant predictor of its impact on strategic stability than its technical capacity alone.

As explained in chapter 8, this problem-set would be exacerbated by the introduction of brittle, inflexible, and opaque AI systems into a situation that disrupted information flows, making it more difficult to assess an adversary's intentions reliably – especially if these intentions alter. Further, the premature deployment of poorly conceptualized and unsafe AI could produce uncertainty and unpredictability, causing a state to either overstate or understate the strategic impact of AI-enhanced capabilities. In sum, to undermine strategic stability, AI would only need to create the perception that credible retaliation at some level of conflict could no longer be assumed as ironclad.

Third, the aggressive pursuit of AI by great powers (especially China and the US) will likely compound the destabilizing effects of AI in a military context. As chapter 4 explained, an increasingly competitive and contested geopolitical world order could mean that the potential military advantages offered by AI-augmented capabilities prove irresistible to states in order to sustain or capture the technological upper hand in the development of frontier (and especially dual-use) technologies vis-à-vis strategic rivals. To be sure, China's progress in multiple military applications of AI merits continued scholarly attention and scrutiny.[3]

Finally, and related, against an inopportune geopolitical backdrop, coupled with the enticing tactical, operational, and strategic perceived benefits of AI-powered weapons (especially AI and autonomy), is the immediate risk posed to strategic stability in the early adoption of unsafe, unverified, and unreliable AI in the context of nuclear weapons, which could have catastrophic implications. For example, fusing AI algorithms with nuclear early warning systems could compress the decision-making timeframe, make the systems themselves vulnerable to manipulation or subversion, and impel states to put their nuclear weapons on high alert status. As demonstrated in chapters 4 and 8, automating (or pre-delegating) escalation would likely worsen crisis stability and increase the risk of inadvertent escalation between great powers.

Part I of the book provided a theoretical and technical framework for conceptualizing the possible implications of AI for strategic stability in the second nuclear age. Chapter 1 defined and categorized the current state of AI (and AI-enabling technologies) and the various limitations of AI, which may prevent it from reliably augmenting autonomous weapon systems – in particular conventional counterforce capabilities. These limitations should encourage caution in the implementation of AI in a military context: a shortage of quality and quantity of data-sets for ML to learn from; the inherent technical limitations of ML algorithms operating in a priori, complex, and adversarial environments; and the vulnerability of these systems to adversarial AI exploitation. The chapter also found that even in the absence of major scientific breakthroughs in AI, incremental steps along the path already established could still have profound positive and negative implications for strategic stability.

Chapter 2 conceptualized AI and technological change in the context of nuclear weapons and strategic stability. It found that strategic stability is ultimately a product of a complex interplay of myriad variables and that technology (including AI) is best viewed as a single variable in the equation. The technology variable can have both stabilizing and destabilizing effects, and the history of emerging technology and revolution in military affairs counsels caution in extrapolating from current (or past) technological trends. Thus, even if AI ultimately has a transformative strategic impact, its impact for the foreseeable future will likely be far more non-linear, iterative, and prosaic. The chapter found that military AI is best understood as a natural manifestation (not the cause) of an established trend in emerging technologies (co-mingling of nuclear and non-nuclear domains, speed of warfare, and compression of the decision-making timeframe), which might persuade (or impel) states to adopt destabilizing nuclear force postures.

Part II of the book considered the strategic competition between China and the US playing out in AI – and other strategically critical technologies – and the possible implications of these developments for US–China crisis stability, arms races, escalation, and deterrence. Chapter 3 looked at the forces driving Washington and Beijing to pursue AI aggressively, and analyzed the impact of the rapid advances and diffusion of AI on the strategic balance. The chapter considered the intensity of US–China strategic competition playing out within a broad range of AI and AI-enabling technologies, in particular Washington's responses to the perceived threat to its technological leadership.

The chapter made three substantive findings. First, a consensus is building within the US defense community that the impact of AI on the future distribution of power, and the military balance, will likely be transformational, if not revolutionary. Second, the proliferation of military AI exists concomitant with a growing sense that the US has dropped the ball in the development of disruptive technologies. Finally, several coalescing features of AI will likely constrain the proliferation and diffusion of AI with militaries' advanced weapon systems, which could harbinger resurgent bipolar strategic competition.

Chapter 4, building on chapter 3, considered how AI-augmented technology might increase the risk of US–China military escalation. It argued that divergent US–China views on the escalation risks of co-mingling nuclear and non-nuclear military technologies would exacerbate the destabilizing effects caused by the integration of these capabilities with AI. It found that if AI-enhanced ISR enabled US counterforce capabilities to more precisely locate China's nuclear assets, this confidence could be undermined, causing destabilizing use-them-or-lose-them situations. Worryingly, there appears to be a lack of understanding (or even recognition) within China's strategic community of the concept of cross-domain deterrence, or how to manage the escalation risks associated with the People's Liberation Army's co-mingled capabilities.

In sum, the multifaceted interplay of AI with both nuclear and conventional weapon systems, deterrence, and escalation, means that a deeper appreciation of this relationship – especially how adversaries view these dynamics – and the implications for crisis decision-making, is fast becoming a critical task for policy-makers.[4] Given the multifaceted interplay of AI with strategic weapons – both nuclear and conventional weapons with strategic effects – a deep appreciation of the dynamic interplay between AI technology, escalation, and crisis decision-making is needed.

Part III of the book provided systematic case studies to consider the escalation risks associated with AI. The chapters in this section explained how and why military AI systems fused with advanced strategic non-nuclear weapons – especially conventional counterforce capabilities – might cause or exacerbate escalation risks in future warfare. They also ruminated on how these AI-augmented capabilities would likely function in practice, and why, despite the risks, great powers will likely deploy them nonetheless.

Chapter 5 considered the impact of AI-enhanced intelligence gathering and analysis systems on the survivability and credibility of states' nuclear deterrence. The chapter found that AI-augmented capabilities (i.e. ISR and remote sensing technology) could make hunting for mobile nuclear arsenals faster, cheaper, and more effective than before. Qualitative improvements to military technology in the nuclear domain will affect states differently, however. This case study also found that irrespective of whether future breakthroughs in AI produce irrefutable evidence of a game-changing means to locate, target, and destroy mobile missile forces, the intentions behind the pursuit of these capabilities by an adversary would be equally (if not more) important.

Chapter 6 considered the risks posed to escalation and deterrence from AI-augmented drone swarms and hypersonic weapons. It found that once the remaining technical hurdles to deploying AI-enhanced drone swarms are overcome, this capability will present a compelling interplay of increased range, precision, mass, coordination, intelligence, and speed in a future conflict. The case study also found that the exponential growth in computing performance, coupled with advances in ML, will enable drones in swarms to perform increasingly complex offensive and defensive missions.

AI is also expected to make significant qualitative improvements to long-range precision munitions, including hypersonic weapons. In aggregate, these improvements could shift the balance further towards the penetrability of states' defense systems and, in turn, reduce the credibility of states' nuclear deterrence. In sum, autonomous weapons such as drone swarms, perceived as low risk with ambiguous rules of engagement, and absent a robust legal, normative, or ethical framework, will likely become an increasingly enticing asymmetric *force majeure* to threaten a technologically superior adversary.

Chapter 7 turned to the AI-cyber security nexus and found that AI-infused cyber capabilities may increase the escalation risks caused by the co-mingling of nuclear and non-nuclear weapons, and the increasing speed of warfare. Future iterations of AI-enhanced cyber counterforce capabilities will complicate the cyber-defense challenge and increase the escalation risks posed by offensive cyber capabilities. The case study also demonstrated that AI tools designed to enhance cybersecurity for nuclear forces might simultaneously make cyber-dependent nuclear support systems more susceptible to cyber-attacks.

Paradoxically, technologies like AI, which improves the reliability and speed of information processed could increase the vulnerabilities of these networks. During crisis conditions, these vulnerabilities could create new first-mover advantages that increase escalation risks. The 'explainability' AI problem-set (see chapter 1) could compound these risks. In short, until AI experts unravel some of the unexplainable features of AI, human error and machine error will likely compound one-another, with unpredictable results.

The final case study, chapter 8, examined the notion of human psychology to elucidate how and why commanders might use AI in the strategic decision-making process, and the risks and trade-offs of increasing the role of machines in the nuclear enterprise – especially the impact of synthesizing AI with nuclear command, control, and communications (NC3) systems.[5] Unlikely as it may be that commanders would explicitly delegate launch authority to machines, AI is expected to be more widely used to support decision-making on strategic nuclear issues.[6] This evolution could bring about significant qualitative improvements to the existing nuclear enterprise, which may increase the confidence in nuclear forces and improve strategic stability.

The case study demonstrated that the distinction between the impact of AI at a tactical and a strategic level is not a binary one. This non-binary distinction could risk – perhaps inadvertently – AI-powered decision support systems substituting the role of human critical thinking, empathy, creativity, and intuition in the strategic decision-making process. In the case of NC3 early warning systems, AI synthesis may further compress the decision-making timeframe, creating new network vulnerabilities that could erode stability.

Furthermore, the nefarious use of AI-enhanced fake news, deepfakes, bots, and other malevolent social media campaigns might have destabilizing implications on effective deterrence and military planning. In sum, as emerging technologies, including AI, are superimposed on states' legacy NC3 systems, new types of errors, distortions, and manipulations involving social media appear more likely to occur. This case study raised several important questions that states will likely respond to very differently. Where in the kill-chain should human operators be inserted, and with what kind of decision-making authority? At what point might human decision-making detract from the advantages (speed and precision) derived from AI and autonomy?

## Managing military AI: avoiding a 'race to the bottom' in AI safety

How can incentives be altered to enhance strategic stability? A prominent theme that runs through this book – and central to understanding the potential impact of AI for strategic stability and nuclear security more broadly – is the concern that AI systems operating at machine-speed will push the pace of combat to a point where the actions of machine actions surpass the (cognitive and physical) ability of human decision-makers to control or even comprehend events.

Possible multi-track (i.e. Track 1, 1.5, and Track 2 discussions) policy responses to push back against the threat posed to stability from AI in a multipolar context can be broadly categorized into two categories.[7] First, those that focus on enhancing debate and discussion between researchers, global defense communities, decision-makers, academics, and other political and societal stakeholders. Second, a range of specific policy recommendations for great military powers to negotiate and implement.[8] Success in these endeavors will require all stakeholders to be convinced of the need and the potential mutual benefits of taking steps towards the establishment of a coherent governance architecture to institutionalize and ensure compliance with the design and deployment of AI technology in the military sphere.

As elucidated in chapter 3, however, geopolitical tensions between great military powers to retain (in the case of the US) or capture (in the case of China) the first-mover advantages in the pursuit of AI will likely create incentives *not to cooperate* on these issues.

## Debate and dialogue

To mitigate, or at least manage, the potentially destabilizing effects of military AI, great military powers must carefully coordinate confidence-building efforts to pre-empt some of the risks to stability that have been highlighted throughout the book.[9] In particular, great powers should establish an international framework for governance, norms, regulation, and transparency in the development and deployment of AI-augmented military capabilities.[10] Further, these frameworks will need to encompass not only the present but also potential future developments, in particular what is and is not being baked into AI algorithms (see chapter 1), and how best to temper the public debate from becoming too fixated on killer robots and machine overlords.

Governments will likely face challenges in these efforts for several reasons. First, AI research and development is widely dispersed across geographic locations and is inherently opaque. Second, the potentially destabilizing and accident-prone features of AI applications can be difficult to identify during the development stages by system engineers. Third, the unpredictability of AI may cause a liability-gap if AI acts in unforeseeable ways, creating legal challenges caused by unintentional or unpredicted harm.[11] Several existing frameworks that govern dual-use technologies, such as space law, internet, and aviation standards, might offer some useful insights for the exploration of AI regulation, demonstrating that even in highly contested military domains, international consensus and areas of compromise can be successfully found.[12]

Furthermore, decision-makers must carefully consider the nuanced trade-offs between increasing degrees of complexity, interdependency, and the vulnerabilities that military AI could engender. It is not an immutable fact (or trajectory) that military systems will be imbued with nascent – and potentially accident-prone – iterations of AI. Instead, these decisions will be made by human policy-makers, tasked with reflecting on these trade-offs, and ultimately, implementing these innovations into safety-critical systems such as NC3 system (see chapter 8).

Similar to the cyber domain, resistance to these efforts will likely come from states who worry that in revealing their military AI (especially offensive) capabilities, they could upend the deterrence utility of these tools.[13] The challenge of coordinating and implementing policies like these will require bold and visionary leadership to circumvent the inevitable regional agendas, interdisciplinary resistance, and burgeoning security dilemmas between strategic rivals. Because of the rapid technological change in AI formal treaties associated with arms-control agreements, which require lengthy and complicated negotiation and ratification processes, legal frameworks risk becoming obsolete before they come into effect. The historical record has demonstrated on several occasions that these kinds of challenges facing humanity can be overcome.[14]

Next, the think-tank community, academics, and AI experts should pool their resources to investigate the implications of military AI for a range of potential security scenarios, including: the impact of AI ML bias reflected in AI-augmented weapon systems; implications of dual-use AI systems for co-mingled nuclear and non-nuclear weapons, and cross-domain deterrence;[15] how to prepare for and react to artificial general intelligence; how investments in research and development might affect the offense and defense balance for AI-augmented military systems; and measures to pre-empt and mitigate offensive uses of AI – both by nuclear powers, non-nuclear powers, and non-state entities (see chapters 7 and 8).[16]

The US National Security Commission on Artificial Intelligence is a new bipartisan commission established by the John S. McCain National Defense Authorization Act for Fiscal Year 2019.[17] The Commission's early work represents a rare example of a collaborative effort between academia, civil society organizations, and the private sector, highlighting the opportunities and risks of utilizing AI for national security purposes. Because of the intrinsic dual-use nature of AI, this dialogue should also be expanded to include other stakeholders, such as private-sector AI and cybersecurity experts, the commercial sector, ethicists, philosophers, civil society, and public opinion.[18]

States should also collaborate on dual-use AI research to leverage AI's low-cost and scaling advantages (i.e. autonomous vehicles and robotics).[19] A focus on the safety, testing, and robustness of AI systems is a critical step to mitigate potential vulnerabilities and risks caused by errors, bias, and explainability in uncontrolled and complex environments.[20] For example, the acceleration in the development of dual-use AI technologies could include publishing performance and safety standards for crucial various military AI applications, as well as creating clear guidelines for modifying commercial AI applications for military use.

The extent of AI integration into dual-use systems such as AI, robotics, and cyber, might influence actors' attitude to risk, the offense–defense balance, and perceptions of others' intentions and capabilities, and could have profound implications for strategic deterrence, nuclear stability, and arms control. A recent study investigated historical treatment of dual-use technologies – biological and chemical weapons, space weapons, cryptography, internet governance, and nuclear technology – to derive insights into applications for AI dual-use risk management policies, such as export controls and pre-publication reviews.[21] This analysis demonstrated the immense difficulty of establishing regulatory, legal, and normative frameworks for dual-use technologies. An example is the cautionary tale of ineffective efforts in the late 1990s to regulate cryptographic algorithms and cyber-network security tools through export controls.[22] Furthermore, in situations where public distribution might create new vulnerabilities and worsen security (e.g. 'adversarial AI'), the publication could be restricted to trusted organizations and entities.[23]

Opportunities for engagement often emerge from innovative, multi-level, cross-domain, cross-cultural, and informal initiatives, laying the foundations for cooperation and discussion at a more formal level – perhaps modeled on the Geneva Conventions or aspects of International Humanitarian Law (IHL) – when geopolitical conditions become more conducive.[24] Therefore, it behooves military institutions to anticipate what AI means for their organizations, strategic culture, and future leaders' development. Failure to do so could mean they will join the long line of military forces whose inability (or unwillingness) to anticipate and adapt to technological change has led to a growing gap between capabilities and operational doctrine.[25]

## Arms control in AI?

Can arms control agreements encompass emerging technologies like AI? How might non-proliferation look in the age of AI?[26] During the Cold War-era, most arms control advocates believed that reciprocal reductions in arms reduced the incentives for disarming first strikes, thus promoting strategic stability. Scholarship on arms control and strategic stability has demonstrated that success in these efforts is, in large part, predicated on the ability of states to delineate precisely between weapon platforms.[27] In the context of advanced and, in particular, dual-use technologies, that assumption will be increasingly tested.[28]

The long-standing Nuclear Non-Proliferation Treaty provides a successful case study in global governance that minimized the threat posed by the weaponization of new (i.e. atomic) technologies while enabling the mutual benefits of sharing nuclear technology to strengthen strategic stability. As shown in chapter 3, however, AI is much more broadly diffused than nuclear technology, with the private sector's heavy involvement in the research, development, and application of AI technology. When the lines between dual-use capabilities, and nuclear and non-nuclear are blurred, arms control is much more challenging, and strategic competition and arms racing is more likely to emerge.[29]

Existing arms control frameworks, norms, and the notion of strategic stability, more broadly, will increasingly struggle to assimilate and respond to these fluid and interconnected trends. Because AI is intrinsically dual-use and non-monolithic, future discussions must consider the implications of AI-related and AI-enabling technologies (see chapter 1), both on the battlefield and at a societal level. Another complicating factor is that today there are no common definitions or engineering methodologies to formulate new regulatory or arms control frameworks to ensure the safety and robustness of AI – the so-called 'AI control problem.'[30] For example, AI experts believe that existing tools (e.g. reinforcement learning techniques) cannot resolve the risks posed to humans from AI-augmented autonomous systems.[31] Thus, a different kind of algorithm than currently exists will likely be needed, prioritizing the safety, robustness, and interpretability of complex AI-infused military systems such as NC3.[32]

Whether AI applications in the military domain can be formally verified, for now, remains an unanswered question. The complexity of AI systems, particularly the difficulty of defining their properties for formal verification, makes them less amenable to verification than other types of technology.[33] For example, the US Defense Advanced Research Projects Agency (DARPA)'s Assured Autonomy Program uses ML algorithms to ensure the safety of autonomous cyber-physical systems. Because this program is designed to learn continuously throughout its lifespan, assurance and verification using traditional methods are very challenging.[34]

This challenge is further complicated by the increasingly cross-domain nature of modern deterrence and the asymmetries emerging in both nuclear and non-nuclear strategic arenas, including cyber, hypersonic weapons, space, and AI.[35] These concerns resonated in the 2018 US *Nuclear Posture Review*. This emphasized that the coalescence of geopolitical tensions and emerging technology in the nuclear domain, particularly unanticipated technological breakthroughs in new and existing innovations – especially affecting nuclear command and control – might change the nature of the threats faced by the US and the capabilities needed to counter them.[36]

To improve strategic stability in an era of rapid technological change, great-power strategic competition, and nuclear multipolarity, the formulation of future arms control frameworks will need to reflect these new shifting perspectives. Arms control efforts can no longer be restricted to bilateral engagement. Governments should also explore ways to increase transparency and accountability for AI and national security, such as addressing the implications of deepfakes and lethal autonomous weapons systems.[37] To counter the threat posed by non-state actors using AI-enabled tools such as deepfakes to manipulate, deceive, or otherwise interfere with strategic decision-making systems in misinformation attacks, governments should – in coordination with both allies and adversaries – continue to harden NC3 systems and processes (e.g. deepfake detection software to detect false information).[38]

Towards this end, in 2017, NATO established a Strategic Communications Center of Excellence, which supports the development of best practices designed to raise awareness of the risks of disinformation posed by the nefarious dissemination of misinformation.[39] In the emerging deepfake arms race, the prospects for detection appear bleak, however.[40] According to photo forensic expert, Professor Hany Farid, "we're decades away from having forensic technology that ... [could] conclusively tell a real from a fake."[41] For instance, the US and China could reconvene the currently suspended Strategic Stability Dialogue – possibly including Russia – to explore issues, inter alia: (1) the impact of AI integration with a range of military (including nuclear) capabilities; (2) the potential for 'new era' AI-infused counterforce and autonomous weapons to unhinge nuclear deterrence; (3) measures to mitigate the risks of inadvertent or accidental nuclear escalation; and (4) promoting collaborative research on AI's impact on international security and safety.[42]

Furthermore, it will be critical for all parties to acknowledge, and where possible seek clarification on, divergences in US and Chinese nuclear doctrines, attitudes to escalation and crisis management, and strategic stability (see chapter 4). These discussions might be incorporated into ongoing broader Track 2 and eventually Track 1.5 dialogues and confidence-building measures in this area, such as clarifying states' intentions, exchanging perspectives on emerging military technologies, and identifying areas for possible mutual interest.[43] How impactful these unofficial dialogues are, leading to tangible improvements in security policies, is an open question. Additional research would be helpful to understand under what circumstances decision-makers might trust the information supplied by AI systems compared to traditional sources, and how regime type (i.e. democratic vs. authoritarian) might make leaders more (or less) inclined to trust and rely on AI systems.

The Chinese government has taken nascent steps to promote research and initiatives on the legal and ethical issues related to AI, including the exploration of rules, safety, regulation, and arms control measures to prevent the potentially destabilizing effects of AI.[44] For example, China's 2017 'New Generation Artificial Intelligence Development Plan' explicitly highlights the need to "strengthen research and establish laws, regulations and ethical frameworks on legal, ethical, and social issues related to AI."[45] Collaboration on specific initiatives may create a foundation for improved understanding and transparency, even against the backdrop of US–China geopolitical tensions and strategic mistrust (see chapter 3).[46] Strategic competition between great powers to reap the perceived first-mover advantages of AI will likely become a negative-sum enterprise without resolving the control and safety issues. *In extremis*, the payoff for all parties might be "minus infinity."[47]

Instead of formal arms control measures or other legally binding agreements, examples of possible normative arrangements that great powers may find mutually beneficial include: prohibiting the development of malicious software and cyber capabilities that target an adversary's NC3 systems, and the use of AI and other emerging technologies in the authorization to launch nuclear weapons (see chapter 8).[48] Other measures that may also improve stability include: reducing the number of nuclear weapons; taking arsenals off high-alert (or launch-on-warning) status; separating warheads from delivery systems (or de-matting warheads); shifting to a deterrent-only (or minimum deterrence) force posture, and adopting a no-first-use declaratory policy – as China and India do today.[49] Unlikely as it may be that these agreements could – technically or politically – be verified for formal compliance purposes, a normative framework or understanding would be worthwhile exploring nonetheless.

One of the most laudable efforts to date for adopting rules concerning acceptable behavior in emerging technologies was expounded by the United Nations (UN), per the General Assembly resolutions on the topic. Specifically, the UN expressed general concern that emerging technologies (especially cyber) might be used for nefarious purposes, "inconsistent with the objectives of maintaining international stability and security," and the body proposed an expert panel to consider "possible cooperative measures to address them, including norms, rules, or principles" of states.[50] In 2015, the expert panel articulated a set of core norms to "prevent the proliferation of malicious information and communication technology tools and techniques."[51]

While voluntary and non-binding, this basic framework may potentially serve as a useful tool to inform any future arms control discussions between states on AI. For the reasons described in Part II of this book, the likelihood that Washington and Beijing (or Moscow) would accept international constraints on the use of technology that targets their respective NC3 and other critical infrastructure is remote. Nonetheless, the continued efforts by bodies like the UN, and prominent commercial and national leaders, to discuss and promote such norms (e.g. clarifying red-lines and rules to enable restraint and tacit bargaining in the digital domain) remains of critical importance.[52] Important questions need to be considered: if a machine (e.g. a drone swarm or automatic target recognition system) violates IHL and commits a war crime, for example, who will be held responsible? Is it the systems operator or designer, the civilian authority who gave the order, or the commander who decided to field the machine?[53]

Some observers have defined this problem as the responsibility-accountability gap, and so consider the use of autonomous weapons as fundamentally unethical.[54] Early efforts towards addressing these issues have recently emerged in the public and private sectors. In 2019 the US Senate passed a 'Deepfake Report Act,' requiring the US Secretary of Homeland Security to write an annual report on the use of deepfake technology "to fabricate or manipulate audio, visual, or text content with the *intent to mislead*" (emphasis added).[55] In the private sector, Facebook launched, in 2019, the Deepfake Detection Challenge, a joint effort between industry, academia, and civil society organizations to explore the research related to the detection of facial manipulation.[56] As deepfakes increase in sophistication, research would also be welcome into digital forensics. Further research would be beneficial on how non-state (i.e. terrorists and criminals) and third-party actors' pursuit of AI might threaten the strategic environment of nuclear-armed powers.

## *Best practices and possible benchmarks in cybersecurity*

Best practices that exist in more sophisticated methods for addressing dual-use concerns, such as computer security, might also be applied to AI – for example, the extensive use of red-teaming exercises to enhance network security and establish a robust organization, governance, and operating practices. Further, AI-cyber red-teaming (e.g. DARPA's Cyber Grand Challenge) will enable engineers and operators to better understand the skills needed to execute particular (offensive and defensive) operations and to manage better system vulnerabilities, adversarial exploitation, stress-testing, and social engineering challenges.[57]

Examples of cybersecurity-centered measures that might be explored to pre-empt and mitigate some of the risks posed by AI-augmented cyber-weapons include the following, albeit non-exhaustive, list.[58] First, coordinating AI-simulated war games, red-teaming creative thinking exercises, and creating redundancies (i.e. back-up or fail-safes) in networks to detect errors, fix vulnerabilities, and increase the reliability, safety, and robustness of military systems, in particular NC3 (see chapter 8). Procedures could be established, for example, to enable confidential reporting, and to fix vulnerabilities, subversions, and other kinds of manipulations, detected in AI-augmented networks.

As highlighted in chapter 7, while some types of AI attacks can occur without accompanying cyber-attacks, more robust cyber defenses will likely increase the challenge of executing specific attacks against AI systems such as data-poisoning. These findings may be used to track the proliferation of AI-related incursions, and then countermeasures developed to manage these threats better.[59] However, because of the intrinsic vulnerability of AI data and algorithms to attack, protecting these systems against such attacks will require a new range of tools and strategies.

Second, governments might spearhead efforts to formalize verification methods and protocols, and consider issues such as the extent, under what circumstances, and for what types of AI systems formal verification can be implemented.[60] Other approaches might also be developed to achieve similar goals, for example ML and big-data analysis, and augmented verification methods.[61] In the case of ML – and other non-deterministic, non-linear, high-dimensional, and continually learning systems – where definitional challenges exist, new standards for verification, testing, and evaluation will likely be needed.[62] In a recent report mandated by the US Congress on the Pentagon's AI initiatives, the authors concluded that the current state of AI verification, validation, testing, and evaluation[63] – both at the DoD and the global defense community more widely – is "nowhere close to ensuring the performance and safety of AI applications," in particular, where safety-critical systems (i.e. nuclear forces) are involved.[64]

An encouraging, albeit rare, example of the private sector taking the lead in this effort is the 'Digital Geneva Convention,' a cyber initiative modeled on the existing post-World War II Geneva Conventions, designed to protect civilians from the negative consequences of cyber-attacks, and supported by, amongst others, Microsoft's President, Brad Smith.[65] Another possible approach might be an industry cooperation regime analogous to the one mandated under the Chemical Weapons Convention, whereby manufacturers must know their customers and report suspicious purchases of sizable amounts of items such as fixed-wing drones, quadcopters, and robotic hardware, which could potentially be weaponized.[66]

Third, the global defense communities should actively invest in the development of AI cyber-defense tools (e.g. analyzing classification errors, automatic detection of remote vulnerability scanning, and model extraction improvements), AI-centric secure hardware, and other fail-safe mechanisms and off-ramps (e.g. circuit breakers used by the US Security Exchange Commission in the stock market), to allow for de-escalation and to prevent unintentional or accidental escalation. Moreover, given that multiple extraneous factors will be central to any decision to move a situation up the escalation ladder (primarily political and strategic contexts), these kinds of technical controls and restraints may *not necessarily* reduce the risk of escalation. Further, applying technological controls will confront the issue of biases and assumptions that are pre-programmed (often unwittingly) into states' ML algorithms.[67]

Several under-explored implications related to these recommendations include the following.[68] How useful would existing tools be against vulnerabilities in AI systems? How would these tools be tailored for AI systems across multiple military domains? Is there an equivalent to 'patching' in military AI systems? What kinds of policies might incentivize, and ensure compliance with, meaningful reforms to existing hardware in the military sphere? How effective would off-ramps and firebreaks be in managing the escalation caused by AI? While these questions are challenging, and necessarily speculative for now, answers will likely become more evident as the technology matures.[69]

Historian Melvin Kranzberg in 1985 opined, "technology is neither good nor bad; nor is it neutral."[70] In other words, emerging technology like AI is not value-neutral – and not necessarily rational. Instead, it reflects the bias, cognitive idiosyncrasies, values (i.e. religious and cultural), and assumptions of its human creators. Measures taken to counter AI ML bias focused on optimizing procedural equity and fairness alone (i.e. neutral algorithms), without sufficient attention on the societal context in which these applications function will simply reinforce the status quo.

Because AI systems reflect the values, bias, and assumptions of humans, and thus the societies in which they reside, resolving these problems is as much a political, ethical, and moral issue as a technical one.[71] Whether AI, as a product of deliberate human actions and decisions, proves to be a stabilizing or destabilizing *force majeure* will largely depend on *how* it is implemented and used, including its technical design (i.e. to account for decision-to-action latency times), operator training and culture, human–machine agency interfaces and trade-offs, and doctrinal assimilation.

Cognizant that some states have (or imminently plan to deploy) AI systems, experts generally agree that AI requires further experimentation, testing, and verification before being integrated into lethal weapon systems and their attendant support systems.[72] Because of the potentially profound effect of AI for human autonomy, security, peace, and even survival, AI research must engage closely with the ethical and legal consequences of developments in these systems. As Alan Turing prophetically stated in 1950, "we can see only a short distance ahead, but we can see that much remains to be done."[73]

## Notes

1 See James S. Johnson, "Artificial Intelligence: A Threat to Strategic Stability," *Strategic Studies Quarterly*, 14, 1 (2020), pp. 16–39.
2 Keir A. Lieber and Daryl G. Press, "Why States Won't Give Nuclear Weapons to Terrorists," *International Security*, 38, 1 (Summer 2013), pp. 80–104.
3 For example, some analysts argue that the apparent contradiction between China's aggressive pursuit of AI-empowered autonomous weapons and its cautious attitude about the deployment of these capabilities can be explained by Beijing's attempt to exert pressure on other militaries (notably the US) whose public opinion is considered more sensitive to the controversies of using automated weapons. Elsa B. Kania, "China's Embrace of AI: Enthusiasm and Challenges," *European Council on Foreign Relations*, November 6, 2018, www.ecfr.eu/article/commentary_chinas_embrace_of_ai_enthusiasm_and_chal lenges (accessed December 10, 2019).
4 The concept of 'asymmetry' is not an entirely new practice in arms control. See Richard K. Betts, "Systems for Peace or Causes for War? Collective Security, Arms Control, and the New Europe," *International Security*, 17, 1 (Summer 1992), pp. 5–43.
5 James Johnson, "Delegating Strategic Decision-Making to Machines: Dr. Strangelove Redux?" *Journal of Strategic Studies* (2020), www.tandfonline.com/doi/abs/10.1080/01402390.2020.1759038 (accessed February 5, 2021).
6 US Department of Defense, *Nuclear Posture Review* (Washington DC: US Department of Defense, February 2018), pp. 57–58.
7 See Johnson, "Delegating Strategic Decision-Making to Machines: Dr. Strangelove Redux?"
8 Center for a New American Security, University of Oxford, University of Cambridge, Future of Humanity Institute, OpenAI & Future of Humanity Institute, *The Malicious Use of Artificial Intelligence: Forecasting, Prevention, and Mitigation* (Oxford: Oxford University, February 2018), pp. 51–55.
9 Concerns relating to the security risks posed by emerging technology (especially lethal autonomous weapons) and maintaining meaningful human control, has led to a variety of initiatives, reports, and other explorative efforts including: reports by the International Committee of the Red Cross; the International Committee for Robot Arms Control; the United Nations Institute for Disarmament Research; and the adoption of a framework and guiding principles by the UN Convention on Certain Conventional Weapons.
10 The 2011 Vienna Document on Confidence and Security Building Measures remains one of the foundational sources of transparency, which could be updated to incorporate AI and autonomous weapons. For instance, remotely operated or unmanned combat aerial vehicles could be included in Annex III of the Vienna Document, together with, combat aircraft and helicopters. Other transparency measures in this document that could also be relevant for lethal autonomous weapons include: airbases visits; demonstration of new types of major weapon systems; prior notification and observation of certain military activities. *Vienna*

*Document 2011 on Confidence-and Security-Building Measures* (Vienna: Organization for Security and Co-operation in Europe, 2011), www.osce.org/fsc/86597?download=true (accessed December 10, 2019).

11 Matthew U. Scherer, "Regulating Artificial Intelligence Systems: Risks, Challenges, Competencies, and Strategies," *Harvard Journal of Law and Technology*, 29, 2 (2016), pp. 354–400.

12 Jacob Turner, *Robot Rules: Regulating Artificial Intelligence* (London: Palgrave Macmillan, 2019).

13 During the Cold War, nuclear deterrence worked in large part because the Soviets and Americans both knew they possessed nuclear arsenals to destroy the other, coupled with confidence in the integrity of this capacity to respond to a first strike.

14 Examples include the 1968 NATO conference at Garmisch, which established a consensus around the mounting risks from software systems, and the 1975 National Institutes of Health conference at Asilomar that underscored the risks posed by recombinant DNA research. Peter Naur and Brian Randell (eds), "Software Engineering: Report on a Conference Sponsored by the NATO Science Committee" (Garmisch, Germany, October 7–11, 1968); Sheldon Krimsky, *Genetic Alchemy: The Social History of the Recombinant DNA Controversy* (Cambridge, MA: MIT Press, 1962).

15 For example, see Dima Adamsky, "Cross-Domain Coercion: The Current Russian Art of Strategy," *IFRI Proliferation Paper* 54 (2015), pp. 1–43.

16 AI-related research findings are often not made public due to reasons related to intellectual property and broader national security concerns.

17 In August 2018, Section 1051 of the Fiscal Year 2019 John S. McCain National Defense Authorization Act established the National Security Commission on Artificial Intelligence as an independent Commission: "to consider the methods and means necessary to advance the development of artificial intelligence, machine learning, and associated technologies to comprehensively address the national security and defense needs of the United States," www.nscai.gov/about/about (accessed December 10, 2019).

18 While many public opinion polls have shown very negative views of the notion of autonomous weapons and AI overall, other studies have demonstrated that the level of negativity can vary significantly depending on how questions are asked. Public opinion can also change dramatically over time, as seen with other previously emerging technologies such as the computer, the VCR, and the telephone. Research on the subject also indicated the malleability of public opinion. It is crucial to recognize that different communities and cultures will have varying abilities to adapt (e.g. technological literacy, culture norms, and economic systems), which may pose challenges for implementing security policies in society at large. See Rob Sparrow, "Ethics as a Source of Law: The Martens Clause and Autonomous Weapons," *ICRC Blog*, November 14, 2017, https://blogs.icrc.org/law-and-policy/2017/11/14/ethics-source-law-martens-clause-autonomous-weapons/ (accessed March 10, 2020).

19 The White House's 2016 'Partnership on AI' series of workshops, the 2017

'Beneficial AI' conference in Asilomar, and the 'AI Now' conference series are good examples of this kind of research collaboration.
20 To address the 'explainability' issue, for example, the US DARPA is conducting a five-year research initiative – the Explainable Artificial Intelligence (XAI) program – to produce explainable AI applications, to gain a better understanding of how they work. Specifically, the XAI program aims to improve trust and collaboration between humans and AI systems by developing ways to explain their reasoning in human-understandable terms. David Gunning, "Explainable AI Program Description," *DARPA*, November 4, 2017, www.darpa.mil/attachments/XAIIndustryDay_Final.pptx (accessed December 10, 2019).
21 Greg Allen and Taniel Chan, *Artificial Intelligence and National Security* (Cambridge, MA: Belfer Center for Science and International Affairs, 2017).
22 Karim K. Shehadeh, "The Wassenaar Arrangement and Encryption Exports: An Ineffective Export Control Regime that Compromises United States' Economic Interests," *American University of International Law Review*, 15, 1 (1999), pp. 271–319.
23 AI researchers recently demonstrated how exposing information about ML algorithms can make them more vulnerable to attacks. See Reza Shokri, Martin Strobel, and Yair Zick, "Privacy Risks of Explaining Machine Learning Models," *arXiv*, December 4, 2019, https://arxiv.org/pdf/1907.00164.pdf (accessed March 10, 2020).
24 By early 2020, at least twenty governments, including the US, China, South Korea, Singapore, Japan, the UAE, the UK, France, Mexico, Germany, and Australia, have actively explored AI policies, initiatives, and strategies of various kinds. Besides, numerous efforts towards intergovernmental cooperation have emerged. For example, in June 2018, leaders of the G7 Summit committed to the "Charlevoix Common Vision for the Future of Artificial Intelligence" to mitigate AI arms racing dynamics www.international.gc.ca/world-monde/international_relations-relations_internationales/g7/documents/2018-06-09-artificial-intelligence-artificielle.aspx?lang=eng (accessed December 10, 2019).
25 For seminal work on this idea see Eliot A. Cohen and John Gooch, *Military Misfortunes: The Anatomy of Failure in War* (New York: Vintage Books, 1990); and Stephen Rosen, *Winning the Next War: Innovation and the Modern Military* (Ithaca, NY: Cornell University Press, 1994).
26 For recent scholarship on the opportunities and pitfalls of efforts to prevent or contain the militarization of AI see Matthijs M. Maas, "How Viable is International Arms Control for Military Artificial Intelligence? Three Lessons from Nuclear Weapons," *Contemporary Security Policy*, 40, 3 (2019), pp. 285–311.
27 Arms control can contribute to arms race stability in three ways: (1) placing reciprocal limits on capabilities; (3) increasing transparency into an adversary's capabilities; and (3) reducing the likelihood of success in the event of military adventurism. Thomas C. Schelling and Morton H. Halperin, *Strategy and Arms Control* (New York: Twentieth Century Fund 1961).

28 Keir A. Lieber and Daryl G. Press, "The New Era of Counterforce: Technological Change and the Future of Nuclear Deterrence," *International Security*, 41, 4 (2017), pp. 9–49.
29 Heather Williams, "Asymmetric Arms Control and Strategic Stability: Scenarios for Limiting Hypersonic Glide Vehicles," *Journal of Strategic Studies*, 42, 6 (2019), pp. 789–813.
30 The AI 'control problem' refers to the issue that, under certain conditions, AI systems can learn in unexpected and counterintuitive ways that do not necessarily align with engineers' and operators' goals. See Stuart Russell, *Human Compatible: Artificial Intelligence and the Problem of Control* (New York: Viking Press, 2019), p. 251; and Joel Lehman et al., "The Surprising Creativity of Digital Evolution," *arXiv*, August 14, 2018, https://arxiv.org/pdf/1803.03453.pdf (accessed March 10, 2020).
31 Dylan Hadfield-Menell et al., "Cooperative Inverse Reinforcement Learning," 30th Conference on Neural Information Processing Systems (2016), Barcelona, Spain, https://arxiv.org/pdf/1606.03137.pdf (accessed March 10, 2020).
32 Jan Leike et al., "AI Safety Gridlocks," *arXiv*, November 27, 2017, https://arxiv.org/abs/1711.09883 (accessed April 10, 2018).
33 Kathleen Fisher, "Using Formal Methods to Enable More Secure Vehicles: DARPA's HACMS program," ICFP 2014: Proceedings of the 19th ACM SIGPLAN International Conference on Functional Programming.
34 Sandeep Neema, "Assured Autonomy," *DARPA*, 2017, www.darpa.mil/program/assured-autonomy (accessed December 10, 2019).
35 Historical examples of asymmetric negotiations and engagement across military domains, including nuclear weapons, are relatively limited, an exception being an option for such facilitation during the negotiation of the INF Treaty that was ultimately abandoned. See Jack Snyder, "Limiting Offensive Conventional Forces: Soviet Proposals and Western Options," *International Security*, 12, 4 (1988), pp. 48–77, pp. 65–66.
36 US Department of Defense, *Nuclear Posture Review* (Washington, DC: US Department of Defense, February 2018), p. 14.
37 Joshua New, "Why the United States Needs a National Artificial Intelligence Strategy and What it Should Look Like," ITIF December 4, 2018, https://itif.org/publications/2018/12/04/why-united-states-needs-national-artificial-intelligence-strategy-and-what (accessed March 10, 2020).
38 For example, Matt Turek, "Semantic Forensics (SemaFor) Proposers Day," *Defense Advanced Research Projects Agency*, August 28, 2019.
39 "NATO Takes Aim at Disinformation Campaigns," *NPR Morning Edition*, May 10, 2017, www.npr.org/2017/05/10/527720078/nato-takes-aim-at-disinformation-campaigns (accessed December 10, 2019).
40 Will Knight, "The US Military is Funding an Effort to Catch Deepfakes and Other AI Trickery," *MIT Technology Review*, May 23, 2018, www.technologyreview.com/s/611146/the-usmilitary-is-funding-an-effort-to-catch-deepfakes-and-other-ai-trickery (accessed December 10, 2019).

41 See Louise Matsakis, "The FTC is Officially Investigating Facebook's Data Practices," *Wired*, March 26, 2018, www.wired.com/story/ftc-facebook-data-privacy-investigation (accessed April 10, 2018).

42 Chinese internal debates have considered the risk of AI and autonomy sparking an arms race or causing unintentional escalation. At the political level, however, efforts to regulate the use of military AI internationally have been viewed as a form of propaganda. For example, see Yuan Yi, "The Development of Military Intelligentization Calls for Related International Rules," *PLA Daily*, October 16, 2019, http://military.workercn.cn/32824/201910/16/191016085645085.shtml (accessed December 10, 2019).

43 For example, see Michael O. Wheeler, "Track 1.5/2 Security Dialogues with China: Nuclear Lessons Learned," No. IDA-P-5135, *Institute for Defense Analyses*, 2014, https://apps.dtic.mil/dtic/tr/fulltext/u2/a622481.pdf (accessed April 10, 2018); and "Track 1.5 US–China Cyber Security Dialogue," *CSIS*, www.csis.org/programs/technology-policyprogram/cybersecurity-and-governance/other-projects-cybersecurity/track-1 (accessed December 10, 2019).

44 Much of China's AI-related initiatives focus on the impact on social stability and the security of the regime against potential internal threats to its legitimacy. The Chinese government is known to routinely collect substantial information on its citizens, regardless of many public statements to the contrary. Gregory C. Allen, "Understanding China's AI Strategy," Center for a New American Security, February 6, 2019, www.cnas.org/publications/reports/understanding-chinas-ai-strategy (accessed March 10, 2020).

45 The State Council Information Office of the People's Republic of China, "State Council Notice on the Issuance of the New Generation AI Development Plan," July 20, 2017, www.gov.cn/zhengce/content/2017-07/20/content_5211996.htm (accessed March 10, 2020).

46 US–China continuous and long-term engagement in scientific and technological collaboration in areas such as global health and climate change demonstrates the potential for mutually beneficial policy issues. See Jennifer Bouey, "Implications of US–China Collaborations on Global Health Issues," Testimony presented before the US–China Economic and Security Review Commission, July 31, 2019, www.uscc.gov/sites/default/files/Bouey%20Written%20Statement.pdf (accessed December 10, 2019).

47 Russell, *Human Compatible*, p. 183.

48 See Robert E. Berls and Leon Ratz, "Rising Nuclear Dangers: Steps to Reduce Risks in the Euro-Atlantic Region," NTI Paper, Nuclear Threat Initiative, December 2016, https://media.nti.org/documents/NTI_Rising_Nuclear_Dangers_Paper_FINAL_12-5-16.pdf (accessed March 10, 2020); Page O. Stoutland and Samantha Pitts-Kiefer, "Nuclear Weapons in the New Cyber Age: Report of the Cyber-Nuclear Weapons Study Group," *Nuclear Threat Initiative*, September 2018, https://media.nti.org/documents/Cyber_report_finalsmall.pdf (accessed March 10, 2020).

49 National Security Commission on Artificial Intelligence Interim Report to

Congress, November 2019, www.nscai.gov/reports, p. 46 (accessed December 10, 2019).
50 UN General Assembly, "Group of Governmental Experts on Developments in the Field of Information and Telecommunications in the Context of International Security: Note by the Secretary-General," A/68/98, June 24, 2013, https://digitallibrary.un.org/record/799853?ln=en (accessed March 10, 2020).
51 UN General Assembly, "Group of Governmental Experts on Developments in the Field of Information and Telecommunications in the Context of International Security: Note by the Secretary-General," A/70/174, July 22, 2015, https://digitallibrary.un.org/record/799853?ln=en (accessed December 10, 2019).
52 A potential precedent for such high-level agreement is the 2015 US–China agreement that prohibited the use of cyberspace for the theft of intellectual property. Notwithstanding the controversy about China's adherence to the agreement, it was generally accepted that Chinese cyber espionage in the US subsided in the immediate aftermath. Adam Segal, "Is China Still Stealing Western Intellectual Property?" *Council on Foreign Relations*, September 26, 2018, www.cfr.org/blog/china-still-stealing-western-intellectual-property (accessed December 10, 2019).
53 IHL's core objectives guiding the use of military force include distinguishing between combatants and civilians, only based on military necessity; is it proportional in terms of the military gains versus the cost imposed on civilians? It uses all practical precautions to help avoid tragedy.
54 Sparrow, "Ethics as a Source of Law: The Martens Clause and Autonomous Weapons."
55 116th Congress, 1st Session, S. 2065 "Deepfake Report Act of 2019," United States Government Publishing Office, October 24, 2019, www.congress.gov/116/bills/s2065/BILLS-116s2065es.pdf. p. 2 (accessed December 10, 2019).
56 The Deepfake Detection Challenge, *Facebook*, 2019, https://deepfakedetectionchallenge.ai (accessed March 10, 2020).
57 See Hyrum Anderson, Jonathan Woodbridge, and Bobby Filar, "DeepDGA: Adversarially-Tuned Domain Generation and Detection," *arXiv*, October 6, 2016, https://arxiv.org/abs/1610.01969 (accessed December 10, 2019).
58 For a recent in-depth review of the social science literature on cybersecurity norms see Martha Finnemore, and Duncan B. Hollis, "Constructing Norms for Global Cybersecurity," *American Journal of International Law*, 110, 3 (2016), pp. 425–479.
59 The technical and practical feasibility of any AI-related security-enhancing hardware or software would also need to be considered by programmers and users.
60 Verification methods may include those used by auditors and regulators to assess AI implementation, such as data science methods, algorithmic design, and robotics hardware. For example, see Aleksandra Mojsilovic, "Introducing AI Fairness 360," September 19, 2018, www.research.ibm.com/artificial-intelligence/trusted-ai/ (accessed December 10, 2019).
61 Center for a New American Security, University of Oxford, University of

Cambridge, Future of Humanity Institute, OpenAI & Future of Humanity Institute, *The Malicious Use of Artificial Intelligence: Forecasting, Prevention, and Mitigation*, pp. 53–54.
62 For an example of new and innovative work on non-linear verification, see Chongli Qin et al., "Verification of Non-Linear Specifications for Neural Networks," *arXiv*, February 25, 2019, https://arxiv.org/pdf/1902.09592.pdf (accessed March 10, 2020).
63 When combined, verification, validation, testing, and evaluation can have essential implications for certification and accreditation of AI technologies and AI-enabled systems. Danielle C. Tarraf et al., *The Department of Defense Posture for Artificial Intelligence: Assessment and Recommendations* (Santa Monica, CA: RAND Corporation, 2019), www.rand.org/pubs/research_reports/RR4229.html, p. 36 (accessed December 10, 2019).
64 Ibid.
65 Cyberspace is not entirely analogous with the Geneva Conventions, however. The Geneva Conventions applied to wartime rules, whereas in cyberspace, standards, regulations, and norms are also needed during peacetime.
66 Ronald Arkin et al., "Autonomous Weapon Systems: A Road-mapping Exercise," *IEEE Spectrum*, September 9, 2019, www.cc.gatech.edu/ai/robot-lab/online-publications/AWS.pdf (accessed December 10, 2019).
67 To protect US consumers, the US Senate recently proposed a new bill that would require companies to audit their ML systems for 'bias and discrimination,' and take corrective action promptly if such issues were identified. 116th Congress, 1st Session, S. 2065 "The Algorithmic Accountability Act of 2019," United States Government Publishing Office, April 10, 2019, www.wyden.senate.gov/imo/media/doc/Algorithmic%20Accountability%20Act%20of%202019%20Bill%20Text.pdf?utm_campaign=the_algorithm.unpaid.engagement&utm_source=hs_email&utm_medium=email&_hsenc=p2ANqtz-___QLmnG4HQ1A-IfP95UcTpIXuMGTCsRP6yF2OjyXHH-66cuuwpXO5teWKx1dOdk-xB0b9 (accessed December 10, 2019).
68 Center for a New American Security, University of Oxford, University of Cambridge, Future of Humanity Institute, OpenAI & Future of Humanity Institute, *The Malicious Use of Artificial Intelligence: Forecasting, Prevention, and Mitigation*, p. 86.
69 Historically, several emerging technologies (e.g. the spread of light water reactors or remote sensing methods) have provided incentives for nuclear energy-aspirants to reveal motives and verify peaceful commitments to lower the risk of conflict, thereby reducing arms racing incentives. Tristan A. Volpe, "Dual-Use Distinguishability: How 3D-Printing Shapes the Security Dilemma for Nuclear Programs," *Journal of Strategic Studies*, 42, 6 (2019), pp. 814–840, p. 816.
70 Gideon Lichfield, "Tech Journalism shouldn't just explain the technology, but seek to make it more of a force for good – Letter from the Editor," *MIT Technology Review*, June 27, 2018, www.technologyreview.com/s/611476/mit-technology-reviews-new-design-and-new-mission/ (accessed December 10, 2019).

71 Fu Wanjuan, Yang Wenzhe, and Xu Chunlei "Intelligent Warfare, Where Does it Not Change?" *PLA Daily*, January 14, 2020, www.81.cn/jfjbmap/content/2020-01/14/content_252163.htm (accessed December 10, 2019).
72 Paige Gasser, Rafael Loss, and Andrew Reddie, "Assessing the Strategic Effects of Artificial Intelligence – Workshop Summary," Center for Global Security Research (Livermore, CA: Lawrence Livermore National Laboratory), September 2018, https://cgsr.llnl.gov/content/assets/docs/Final_AI_Workshop_Summary.pdf (accessed December 10, 2019).
73 Quoted in Stuart Russell and Peter Norvig, *Artificial Intelligence: A Modern Approach*, 3rd ed. (Harlow: Pearson Education, 2014), p. 1067.

# Index

*Note*: Page numbers followed by 'n' and note number refer information in chapter notes. Page numbers in italic type refer to figures.

5G networks 63, 65–66, 68, 71, 72, 77n32, 178, 180, 184

accidental escalation *see* inadvertent escalation
Acton, James 136
advanced persistent threat (APT) operations 153, 156, 159
AeroVironment 144n40
AGI *see* artificial general intelligence (AGI)
Allen, John 69
AlphaGo system 2, 62, 170, 173, 190n48
AlphaStar system 170
anti-submarine warfare (ASW) 7, 112, 115–119, 119–120, 121n13, 132, 133, 134, 153
APT operations *see* advanced persistent threat (APT) operations
arms control 27, 41, 44, 131, 146n57, 205, 206–212
arms race stability 43, 44, 45–46, 63, 64, 215n27
artificial general intelligence (AGI) 18–19, 83n91, 205
artificial intelligence (AI)
  central theoretical framework 6, 42, 47–48, 49, 200
  current state of AI and AI-enabled technology 5, 17, 18–25, 30–31, 200

definitions 19–21, 74n1
demystifying the hype 6, 17, 26–30, 31
intergovernmental cooperation 41, 131, 146n57, 203–212
limitations 17, 21, 23–25, 30–31, 173, 175, 200
  *see also* explainability problem
machine learning as a critical enabler 17, 21–25, 139n2
in science and engineering 5, 17, 18, 19, *19*, 47
ASW *see* anti-submarine warfare (ASW)
ATR systems *see* automatic target recognition (ATR) systems
Australia 69, 79n44, 215n24
automatic target recognition (ATR) systems
  cyber attacks on 154, 159
  drone swarms 133
  hunting for nuclear weapons 38n73, 112, 113, 114, 115, 198
  ML limitations 24, 25
automation bias 158, 170–171, 175–176, 182, 185
autonomous sensor platforms 29, 38n65, 112, 115, 129, 133, 153–154
autonomous strategic decision-making 8, 202–203
  advantages of 28, 39n81, 177–178

autonomous strategic decision-making (*cont.*)
  agreed usage 130, 170
  alarmist fears 170
  data limitations 173, 175
  escalation risks 91, 169, 173–176, 178–183, 184–185, 200, 203
  games 170
  human psychology 28, 170–171, 175–176, 182
  and the non-binary tactical and strategic distinction 8, 173–175, 183–184, 203
  regime type 28, 169, 171–172, 181, 208
autonomous vehicles 29–30
  *see also* drones; hypersonic weapons; unmanned aerial vehicles (UAVs); unmanned surface vehicles (USVs); unmanned underwater vehicles (UUVs)
autonomy 20, 20, 22, 32n3, 34n22, 120–121n5

B-21 *Raider* 144n42
Baidu 67
ballistic missiles 46, 51n17, 99n11, 121n8, 126n58, 133, 135, 136, 168
  *see also* intercontinental ballistic missiles (ICBMs); nuclear-powered ballistic missile submarine (SSBNs); submarine launched ballistic missile (SLBMs)
bandwagoning 44–45
bargaining situations 44, 45, 88, 92, 190n51
big-data analytics 18, 26, 32n7, 137
  early warning systems 174, 178, 180, 182, 184
  nuclear deterrence utility 7, 111, 112, 114, 116, 119, 120, 121n9
bipolarity 60, 61, 64, 70, 71, 73, 201
black box problem *see* explainability problem
Brodie, Bernard 47, 150

C3I (command, control, communications, and intelligence) systems
  cyber vulnerabilities 53–54n45, 152, 153, 154, 155, 158, 159
  drone attacks 131, 133, 135
  entanglement problems 53–54n45, 86, 173
  US–China crisis stability 93, 114, 118, 135
Carter, Ashton 61
chemical weapons 47
China
  autonomous vehicles 67, 118–119, 123n32, 130, 133–134, 141n12
  hypersonic weapons 38n63, 92, 118, 135, 136, 137, 138
  internal security 39n76, 193n72, 217n44
  international dialogue and arms control 54n53, 130, 146n57, 204, 208–209, 215n24, 218n52
  limited nuclear resolve 92, 194n79
  machine-based decisions 130, 171–172, 174, 176
  submarines 117, 118, 127n66
  and the US-China technology gap 2, 6–7, 59–60, 60–69, 71, 72, 73, 200–201
  *see also* US–China crisis stability
civilian AI 18, 63, 207
  and arms-racing dynamics 69–70
  China's civil-military fusion agenda 2, 62, 64, 65–67, 68–69, 71, 194n86
  'Digital Geneva Convention' 211
  Pentagon-Silicon Valley collaboration 3, 61, 67
Clapper, James 153
Coats, Daniel 63, 150
Cold War
  arms control 206
  automated early warning systems 168, 188n29, 193n73
  bipolarity 46, 69–70
  Cuban missile crisis 194n82

# Index

escalation dynamics 89, 90, 99n13, 102n44, 102n51, 104n67, 104n68, 164n43
*Sputnik 1* 79n19
strategic stability 41, 42, 45–46, 48, 98n5, 214n13
target intelligence 38n73
US counterforce capabilities 101n36
co-mingling problem 43, 45, 85, 98n2, 102n45
cyber capabilities 86, 150, 151, 202
drone swarms 133
NC3 systems 53n45, 173
trends exacerbating escalation risk 53–54n45, 86–87
US-China crisis stability 7, 85, 86, 89–90, 92–94, 96, 97, 201
warhead ambiguity 87, 136, 194n82
command and control systems 29
accident prevention 122n20
autonomy 28, 168, 170, 172, 179, 183
cyber attacks 27, 152, 155, 156
drones 139n1, 143n29
locating mobile missiles 121n7
quantum communication 194n86
US–China crisis stability 87, 90, 105n75
*see also* C3I (command, control, communications, and intelligence) systems; NC3 systems
commercial AI *see* civilian AI
computing power 2, 18, 22, 70, 71, 138, 178, 202
Cote, Owen 115
crisis stability 43–44, 46, 47–48, 49, 86–89
AI-augmented drones 8, 118, 128, 129, 131, 138
asymmetric information situations 88, 89, 157, 158, 159, 176, 179
autonomous strategic decision-making 169, 173–176, 178–183, 184–185, 200, 203
early warning systems 169, 184, 199–200, 203
psychological factors 44

*see also* escalatory risks of AI; US–China crisis stability
cruise missiles 12n34, 92, 118, 129, 135, 136, 147n65, 148n79
Cruz, Ted 63
cyber warfare
AI-cyber security nexus 8, 150–159, 202
data-poisoning 153, 155, 178, 211
drone swarms 47, 129–130, 133, 134, 138, 151
NC3 systems 27, 53–54n45, 88, 152–153, 154, 155, 156, 162n22, 176, 177, 185
regulatory, legal and normative frameworks 206, 207, 208, 209, 210–212, 218n52
*see also* deepfakes; false-flag operations; spoofing attacks

DARPA *see* Defense Advanced Research Projects Agency (DARPA, US DoD)
data
limiting military AI 23–25, 83n92, 173, 175, 200
non-military AI 22
for nuclear domain 173, 175
'poisoned data' attacks 153, 155, 178, 211
*see also* big-data analytics
debate and dialogue 9n4, 26, 131, 203–206, 208
decision-making *see* autonomous strategic decision-making
deepfakes 157–158, 178–179, 180, 181, 185, 208, 210
deep learning (DL) 11n20, 21, 22, 23, 111, 148n78, 160n6, 174, 191n55
DeepMind 2, 76n19, 81n62, 170, 187n12
Defense Advanced Research Projects Agency (DARPA, US DoD) 28, 174, 177–178, 186n6, 207, 210, 215n20
*see also* Sea Hunter

deliberate escalation *see* intentional escalation
deterrence 100n23
  AI-enhanced cyber capabilities 151, 152–153, 155, 157
  arms control 207, 208, 209
  autonomous decision-making 170, 175, 179
  Cold War 98n5, 102n44, 214n13
  co-mingling problem 102n45
  drone swarms 8, 91, 117, 128, 131, 132, 136, 138, 202
  hypersonic weapons 8, 92, 128, 135, 138, 202
  information flow 88, 89, 96, 157, 158, 159, 176, 179, 180
  international dialogue 205
  Korean War 103n56
  malevolent social media campaigns 180, 185, 203
  mobile missiles 111, 114, 119, 198
  in multipolar nuclear order 95
  NC3 systems 168–169, 189n36
  nuclear submarines 111, 112, 116, 117, 118, 120, 132, 198
  perception vs, technical capabilities 4, 199
  in strategic stability framework 41, 44, 45, 47, 85
  survivable nuclear arsenals (or second-strike capabilities) 45
  US–China crisis escalation framework 73, 85, 86–87, 89, 90, 91, 92, 93, 97, 113–114, 118, 119, 120
  World War I 47
  *see also* first-strike stability
dialogue *see* debate and dialogue
Digital Geneva Convention 211
DL *see* deep learning (DL)
drones 8, 27, 30, 47, 113, 128–135, 138–139, 202
  *see also* unmanned aerial vehicles (UAVs); unmanned surface vehicles (USVs); unmanned underwater vehicles (UUVs)
Dunford, Joseph 153

early warning systems 27
  autonomous decision-making 199
  China 136, 148n78, 174, 179
  Cold War 168
  cyber vulnerability 152, 153, 154, 158, 166n65, 178–179
  deterrence 114
  drone attacks 131, 133
  entanglement problem 86, 90
  hypersonic attacks 135–136
  pre-emptively attacked by drones 128, 131, 133
  warhead ambiguity 136, 173
  *see also* NC3 systems
entanglement *see* co-mingling problem
escalatory risks of AI 1, 4–5, 111
  AI-cyber security nexus 8, 87, 150–159, 176, 178, 185, 202
  autonomous strategic decision-making 130, 138, 169, 173–176, 178–183, 184–185, 200, 203
  fear and risk aversion 43, 44, 46, 88, 131, 139
  hypersonic missile delivery 8, 118, 135–136
  perception vs. technical capacity 4, 43, 199
  tracking and targeting capabilities 7–8, 111–112, 132
    anti-submarine warfare 116, 117–118, 120, 132
    land-based missiles 112, 113–115, 119, 133
  *see also* co-mingling problem; crisis stability; US–China crisis stability
explainability problem 25, 89, 154, 159, 170, 176, 184, 202, 205

F-35 fighter jets 129, 130, 144n41
false-flag operations 87, 157, 182
Farid, Hany 158, 208
feedback loops 19, 28, 95
first-strike stability 28, 43, 45–46, 113, 184
  AI-enhanced ISR systems 7, 112, 113, 114–115, 198

# Index

anti-submarine warfare 112, 116, 118, 132
and the capability-vulnerability paradox 151, 154, 155, 159
conventional warfighting doctrine 54n45, 86–87, 92, 98
deepfakes 157
drone swarms 118, 131–132
hypersonic weapons 135, 137
launch-on-warning nuclear posture 152, 158–159
vulnerable command and control systems 43, 90, 93, 114, 155, 156, 181–182
force structures 2, 7, 28, 90, 118, 120, 169, 183
France 54n53, 69, 82n75, 141n14, 215n24

GANs (generative adversarial networks) 166n66, 178
Germany 79n44, 215n24
Gilday, Michael 155
Goldstein, Avery 92
Google 2, 38n66, 67, 78n38, 173
Go-playing 2, 62, 170, 173, 190n48
Gottemoeller, Rose 182
Gower, John 117
Gray, Colin 29, 44

Halperin, Morton 45
Heginbotham, Eric 114–115
Huawei 66
Husain, Amir 69
hypersonic weapons 146–147n60, 147n62
China 38n63, 92, 118, 135, 136, 137, 138
and missile defenses 8, 92, 128, 129, 135–138, 167n71, 202
Russia 38n63, 135, 137, 147n68
US 135, 147n68, 148n76, 148n81

ICBMs *see* intercontinental ballistic missiles (ICBMs)
image recognition software 22, 23, 24–25, 29, 154

inadvertent escalation 4–5, 27, 43, 85, 86–87
in AI-cyber security nexus 87, 150, 151, 153–154, 157–158, 159, 176, 178, 185
autonomous strategic decision-making 130, 138, 169, 173–176, 177–183, 184–185, 200, 203
drone swarms 128, 132, 136, 139
hypersonic weapons 135
perception vs. technical capacity 4, 43, 199
US–China crisis stability 7, 86, 87, 89–96, 97, 136, 201
warhead ambiguity 179
*see also* co-mingling problem
India 54n53, 98n7, 107n94, 196n101, 209
intelligence, surveillance, and reconnaissance (ISR) systems 21
anti-missile location 112, 113–115, 119, 131, 133, 137, 198
and anti-submarine warfare 115, 117–118, 134, 198
strategic decision-making 174
swarm combat 117–118, 128, 131, 132, 133, 134, 136, 138, 149n88
technical bottlenecks 23–24
US–China competition 64, 67, 94, 97, 113–114, 115, 135–136, 201
intentional escalation 43, 90, 91, 96, 98n3, 99n17, 107n90, 131, 154
intercontinental ballistic missiles (ICBMs)
autonomy 148n74, 174
and drone swarm technology 37n56, 128, 133, 138
locating and targeting mobile launchers 38n65, 38n73, 112, 120, 132, 198
mission command 192n67
speed of war 186n3
intergovernmental cooperation 41, 131, 146n57, 203–212
IoT (internet of things) 153, 155

Iran 103n61
Iraq 29, 141n11, 163n31
ISIS 134, 141n11, 180
Israel 11n20, 69, 121n6, 140n10
ISR systems *see* intelligence, surveillance, and reconnaissance (ISR) systems

Japan 193n72, 195n95, 215n24
Jervis, Robert 45, 151

KAIROS 173
Kranzberg, Melvin 212

LAMs *see* loitering attack munition (LAMs)
LAWs (lethal autonomous weapons) 130, 131, 134, 139, 146n57, 208, 213n9
Lieber, Keir 45
limited nuclear war 92, 93, 113, 119, 194n79
loitering attack munition (LAMs) 129, 144n40

machine learning (ML) 10n9, 32n2
   algorithmic limitations 24–25, 71, 73, 200
   biases and assumptions 24, 28–29, 212
   C2 processes 170
   as a critical AI-enabler 17, 21–25, 139n2
   cyber warfare 150–151, 152, 153, 154, 155, 157, 159, 176, 177
   data requirements 18, 23–25, 173, 175, 200
   drones 129, 130, 132, 134, 138, 145n44
   explainability problem 25, 89, 154, 159, 170, 176, 184, 202, 205
   Go-playing 2, 170
   hypersonic delivery systems 8, 129, 137
   improving situational awareness 87–88, 94, 174, 177
   locating, tracking and targeting nuclear weapons 7, 111, 112, 113, 116, 117, 118, 120
   and major research fields 19
   military implications (military primer) 5, 17, 29
   NC3 systems 173, 174, 175, 176, 177
   neural networks 25, 26, 30, 32n8, 34n31, 38n68, 118, 166n66, 178
   quantum ML 67, 78n38
   RAID ML algorithm 177–178
   space technology 144n36
   unintended learning 154, 216n30
   and US technical hegemony 26–27, 61, 67, 70, 71
   verification 211
   *see also* deep learning (DL)
Mahan, Alfred 111
Maven, Project 29, 67, 160n6
Mearsheimer, John 95
missiles *see* ballistic missiles; intercontinental ballistic missiles (ICBMs); mobile missiles; nuclear-powered ballistic missile submarine (SSBNs); submarine launched ballistic missile (SLBMs)
mission command 177
mobile missiles
   AI-augmented drone attacks 113, 131, 133, 135, 137, 138
   AI data limitations 24, 25
   AI-enhanced locating 29, 112, 113–115, 119, 123n30, 137, 148–149n82
   eroded deterrence role 111, 112, 114, 119, 198
   hypersonic delivery systems 137
   US–China crisis stability 93, 94, 113, 119
multipolarity 60, 61, 62, 64, 69–71, 73
   *see also* nuclear multipolarity

narrow AI 18
National Security Commission on Artificial Intelligence 205

# Index

NATO 102n44, 208, 214n14
NC3 systems 8, 28, 168–169, 172
   AI-enhanced fake news 180–182, 183, 185, 203, 208
   automated escalation 169, 173, 174, 175–176
   cyber-attacks 88, 152–153, 155, 156, 176, 177, 178–179, 185
   cyber-defenses 27
   entanglement problem 53n45, 173
   intentional escalation scenario 90
   located by drone swarms 138
   regime characteristics and AI incorporation 171–172
neural networks 25, 26, 30, 32n8, 34n31, 38n68, 118, 166n66, 178
North Korea 29, 50n9, 82n75, 98n7, 103n56, 107n94, 165n51, 196n101
nuclear command, control, and communications systems *see* NC3 systems
nuclear doctrine 28, 51n24, 54n53, 90, 104n67, 164n43, 172, 179, 208
nuclear multipolarity 5, 44, 48, 50n15, 71, 95–96, 99n10
Nuclear Non-Proliferation Treaty 206–207
nuclear-powered ballistic missile submarine (SSBNs) 37n56, 112, 115–119, 120, 121n13, 128, 131, 132

offshore balancing 44
*Operation Bodyguard* 194n78

Pakistan 98n7, 103n61, 107n94, 181, 194n79, 196n101
*Perdix* drones 144n38
polarity 6, 59, 61, 74n5
   *see also* bipolarity; multipolarity; nuclear multipolarity
policy recommendations 204, 206–212
   *see also* arms control
Poseidon (Status-6) 144n42, 198n115

predictive policing 28, 193n72
Press, Daryl 45
private sector *see* civilian AI
Project Maven 29, 67, 160n6
Putin, Vladimir 2, 123n26

quantum computing 63, 67, 78n38, 79n51, 116, 126n63, 178, 194n86

RAID (Real-Time Adversarial Intelligence and Decision-Making) ML algorithm 177–178
red-teaming 27, 62, 210
regime type 28, 169, 171–172, 181, 208
regulation 204–206
responsibility-accountability gap 25, 169, 174, 183, 210
revolution in military affairs 2, 3, 6, 19, 47, 49, 199, 200, 201
robotics
   AI and autonomy linkages 20, *20*
   in the AI ecosystems 21, *23*
   Chinese investment 2, 63, 64, 65, 78n40
   industry cooperation 211
   international dialogue 131, 205
   major research fields *19*
   swarming missions 130, 134, 139n1, 140n5, 142n22
   tactical classification 27
   US investment 62, 68
Rose, Frank 85
Russia
   AI cyber-defenses 155
   drone technology 141n13, 142n22, 144n38, 144n43
   escalation dominance 86
   hypersonic weapons 38n63, 135, 137, 147n68
   limited nuclear resolve 194n79
   mobile missiles 29
   and the multipolar world order 107n94
   NC3 system vulnerability 156
   nuclear doctrine 51n24, 54n53, 152, 191–192n62

Russia (*cont.*)
　Soviet Union *see* Cold War
　and US ISR capabilities 114
　and US technological hegemony 61–62

safety
　cyber security 157, 159, 163n36, 210, 211
　management of 203–206, 207, 208, 209
　and US technical hegemony 62, 68–69
　verification and testing 27, 41, 129, 199, 211, 212
satellites 81n69, 106n85, 113, 127n68, 168, 174, 178, 189n35, 196n101
Scharre, Paul 112, 139
Schelling, Thomas 43–44, 45, 91, 175, 180
*Sea Hunter* 113, 116, 132
second-strike capabilities 4, 157, 198–199
　cyber attacks 87, 88
　and first strike stability 43, 45
　ISR enhancement 113
　proclivity for autonomous decision-making 112, 172, 184
　stealthy SSBNs 112, 116, 132
　and US–China stability 92, 93, 96
Second World War 145n52, 194n78, 195n95
security dilemma 64, 87, 96, 118, 139, 160n4
Sedol, Lee 170
Shanahan, Jack 23–24, 63
signaling intent
　AI and pre-programmed mission goals 175, 176
　cyber operations 151, 152, 156, 164n49, 166n62
　fake social media 180–181
　hypersonic weapons 135
　resolve and deterrence 41, 87, 92, 97, 114
　'stability–instability paradox' 95–96
　suggestive escalation 107n91
　Vietnam War 190n50

Silicon Valley 3, 61, 64, 67
Singapore 69, 82n75, 193n72, 215n24
*Skyhawk* 126n64
Smith, Brad 211
social media 180–181, 182, 183, 185, 203, 210
South China Seas 61, 118
South Korea 69, 82n75, 144n38, 180–181, 215n24
Soviet Union *see* Cold War
spoofing attacks 28, 138, 153–154, 159, 195n95
*Sputnik 1* 76n19
SSBNs *see* nuclear-powered ballistic missile submarine (SSBNs)
strategic decision-making *see* autonomous strategic decision-making
strategic stability 1, 198–199, 200
　analytical framework 6, 41–48, 200
　arms control 27, 41, 44, 131, 146n57, 205, 206–212
　debate and dialogue 9n4, 26, 131, 203–206, 208
　qualitative improvements and autonomous strategic decision-making 28, 39n81, 177–178
　*see also* arms-race stability; co-mingling problem; crisis stability; escalatory risks of AI; first-strike stability; US–China crisis stability
submarine launched ballistic missile (SLBMs) 136, 148n74, 186n3
submarines *see* anti-submarine warfare (ASW); nuclear-powered ballistic missile submarine (SSBNs); submarine launched ballistic missile (SLBMs)
swarm combat 8, 27, 128–131, 202
　anti-submarine warfare 115, 116, 117–118, 121–122n13, 134
　China 67, 133–134, 144n38
　cyber warfare 47, 129–130, 133, 134, 138, 151
　new strategic challenges 8, 131–135, 138–139

# Index

Taiwan Straits 61, 118
target ambiguity 96, 135
TELS *see* transporter-erector-launchers (TELs)
terminal guidance capabilities 112, 113, 198
terrorism 29, 134, 141n11, 180, 182, 185, 189n42, 210
testing *see* verification and testing
transporter-erector-launchers (TELs) 121n8, 123n30, 196n103
Turing, Alan 212

UAS (unmanned autonomous systems) *see* drones
UAVs *see* unmanned aerial vehicles (UAVs)
underwater gliders 118, 124n41
unintentional escalation *see* inadvertent escalation
United Kingdom 54n53, 79n44, 82n75, 83n84, 124n35, 141n14, 193n72, 215n24
United Nations 146n57, 209, 210
United States of America
  AI investment 3, 26–27
  anti-submarine warfare 113, 116, 117, 118, 123n32, 127n66, 132, 145n50
  automated decision-making 130, 140n9, 149n86, 168, 174, 177–178
  B-21 *Raider* 144n42
  ballistic missile defense 51n17
  cyber capabilities 27–28, 64–65, 151
  cyber vulnerability 130, 150, 151, 162n22
  debate and dialogue 54n53, 146n57, 205, 208, 215n24
  drone swarm technology 130, 144n39, 144n40
  hypersonic weapons 135, 147n68, 148n76, 148n81
  mobile missile location (including Project Maven) 29, 67, 93, 94, 113–114, 119, 160n6
  nuclear strategy 50n9
  *Perdix* UAV 144n38
  predictive aircraft maintenance 37n60
  predictive policing 193n72
  responsibility-accountability gap 210
  Silicon Valley 3, 61, 64, 67
  systems verification 207
  and the US-China technology gap 3, 6–7, 59–60, 60–69, 71, 72, 73, 200–201
  *see also* Cold War; US–China crisis stability
unmanned aerial vehicles (UAVs)
  anti-submarine warfare 115, 116, 117, 126n55, 134
  battery power 121–122n13, 143n33, 146n56
  China 67, 126n64, 130
  cyber-attacks 129–130, 133, 134, 151
  escalation dynamics 132
  locating mobile missiles 121n10
  non-autonomous 142n20
  nuclear missile delivery 37n56, 128, 133
  Project Maven 29, 67, 160n6
  reducing human error 177
unmanned autonomous systems (UAS) *see* drones
unmanned surface vehicles (USVs) 113, 115, 116, 117, 132, 134
unmanned underwater vehicles (UUVs)
  anti-submarine warfare 115, 116, 117, 118, 134, 136
  battery power 117, 121–122n13, 146n56
  China 67
  nuclear missile delivery 37n56, 128, 133
  Poseidon (Status-6) 144n42, 198n115
  US–China diplomatic incident (2016) 91

US–China crisis stability 6, 7, 73, 96–97, 198, 199, 201
  autonomous vehicles 118, 130
  China's sanguine attitude to inadvertent escalation 7, 85, 92–94, 97, 113, 201
  control and safety issues 208–209, 218n52
  cyber security 86, 87–88, 152, 155, 156, 179
  divergent thinking on escalation risk 7, 85, 86, 87, 89–91, 92–93, 96, 201
  hypersonic weapons 92, 118, 135, 136
  mitigating and managing escalation risk 86, 94–96
  mobile missiles 93, 94, 113–114, 119
  new escalation pathways 64, 86–89, 96, 156
  submarine warfare 118–119, 120

UUVs *see* unmanned underwater vehicles (UUVs)

verification and testing 27, 41, 129, 199, 211, 212
Vienna Document on Confidence and Security Building Measures (2011) 213n10

Walton, Dale 44
wargaming 4, 27, 62, 170, 174, 210
warhead ambiguity 87, 136, 194n82
Work, Roberts 130
world order *see* polarity
World War II 145n52, 194n78, 195n95

Xi Jinping 66

Zhuhai Ziyan 133–134

Printed in the USA
CPSIA information can be obtained
at www.ICGtesting.com
JSHW012351190724
66732JS00004B/30